The Only Sacred Ground

Scientific Materialism and a Sacred View of Nature Within the Framework of Complementarity

The Only Sacred Ground

Scientific Materialism and a Sacred View of Nature Within the Framework of Complementarity

Gregory N. Derry

Department of Physics
Loyola University Maryland

Apprentice House
Baltimore, Maryland

First Edition

Printed in the United States of America

Paperback ISBN: 978-1-62720-020-2
Ebook ISBN: 978-1-62720-021-9

Design by Lauren Andes
Cover illustration by Albrecht Dürer

Published by Apprentice House

Apprentice House
Loyola University Maryland
4501 N. Charles Street
Baltimore, MD 21210
410.617.5265 • 410.617.2198 (fax)
www.ApprenticeHouse.com
info@ApprenticeHouse.com

....in the great drama of existence we are ourselves both actors and spectators.

—Niels Bohr

this is my home, this is my only home, this is the only sacred ground that I have ever known

—David Carter

Contents

Preface...ix

1. Introduction 1

2. Ways of Understanding Nature 19

3. The Mundane World 55

4. The Sacred World..65

5. The Desacralization of Nature 73

6. Quantum Mechanical Background 99

7. Bohr and Complementarity...109

8. Revisioning Complementarity 125

9. Epistemological Questions..139

10. Creation 157

11. Mind and Brain..183

12. A New View of Old Issues..225

Epilogue 257

Endnotes 261

Bibliography..267

Preface

This book is an extended reflection on our ideas about nature, written by a scientist, sympathetic to and informed by science, but very definitely not a scientific treatise. The approach adopted here is a transdisciplinary approach that draws on and respects a host of disciplines (philosophy, history, biology, theology, cultural studies, religious studies, and my own field of physics, among others) in an integrative manner. I do not pretend to have genuine mastery of all these areas, but I think it's important for some people to make the attempt at transdisciplinarity lest the increasing fragmentation of our knowledge, our culture, and our personhood becomes oppressive. I apologize for any mistakes or oversimplifications I may have made, but I don't apologize for making the attempt to integrate a large amount of material from a wide array of disciplines into a coherent synthesis.

Although I became quite interested in the literature on science/religion relationships while writing this, and although I consider the present work to be a contribution to that literature, this book is not a generic book on science and religion. Instead, it is a more detailed look at a very specific idea (complementarity) and how this idea contributes to the solution of a specific problem (the tension between a sacred apprehension of nature and scientific materialism). The major original idea presented here, the heart of the argument, is found in chapter 8. The material preceding chapter 8 is mainly to provide sufficient context to make that chapter intelligible. The material following chapter 8 is mainly to deepen and illustrate the potential applications of the idea to various specific problems.

Here at the outset I would like to clear up two possible misconceptions. The first possible misconception is that I am somehow trying to apply quantum mechanics to religious questions. Although it is true that complementarity has its roots in quantum mechanical issues, the argument I am making here is an epistemological argument that does not employ quantum theory *per se*. The second possible misconception is that I am trying to devise some sort of grand system that will explain everything about nature. My goal is both more modest than this and also more practical: I am attempting to provide an additional methodological tool to analyze and think about the relevant issues.

I started this project more than eight years ago, during a sabbatical year. The project began on a surreal note, as I was sitting outside reading about Niels Bohr's philosophy and the seventeen-year cicadas started to emerge from the ground. Before the year ended, I was locked in a battle with a stage-three lymphoma, which I ultimately survived. Writing the book has lead to a number of other interests and to my encounters with a host of interesting thinkers. The present publication of the book is a gratifying end to a long journey, and I'd like to thank Apprentice House for their decision to publish it.

I also have a number of other debts to acknowledge. First, I would like to thank the John Templeton Foundation for a grant to work on the book during my sabbatical. I would also like to thank my institution, Loyola University Maryland, for providing the sabbatical year and supporting my work. Although I have had many stimulating conversations about this work with many different people, I would particularly like to single out Nicholas von Stillfried and Kris Jargocki in this respect. I'd also especially like to thank K. Helmut Reich for a thorough reading of the entire manuscript and a huge number of constructively critical comments. Closer to home, this work has benefited immeasurably from my intellectual interactions with my wife, Paula Derry, my daughter, Rebecca Derry, and my friends Daniel Perrine, Richard Blum, and Lisa Blum. And although I have never met or talked to Jan Faye or Henry Folse, I owe both of them a great deal for their excellent Bohr scholarship, which I have exploited extensively here. Lastly, I would like to thank Tracy Grammer for allowing me to use the fragment of song lyrics by the late poetic genius Dave Carter in the epigraph; these lines gave me the inspiration for the title of the book.

Although in some ways this is an entirely intellectual work (an exercise in logical analysis, as it were), on another level it has the deeper purpose of trying to create a space for spiritual realities in the modern world in a manner consistent with reason. My hope is that someone may find reading it to be useful in their own struggles to do this in their lives.

Baltimore
October 31, 2013

1. Introduction

Each of us apprehends the world through our lived experience. Our senses and our reasoning faculties (not to mention our emotional states, memories, etc.) present us with the world as we directly know it. Is this the way the world "really is" or is there some underlying reality behind the veil? If so, is this underlying reality knowable or not? Or is there really nothing real there at all, with only our own constructions left over, to which we impute more substance than they deserve? These are perennial questions, which have been asked repeatedly for thousands of years. But in the last several hundred years, in western culture, we have had available a new tool with which to explore these issues: scientific thinking. Using this tool, we've learned a fantastic amount about the inner workings of nature, and science has indeed penetrated several layers behind the veil of our direct perceptions and common-sense thought patterns. Long before science as we understand it was conceived, however, humans employed ancient methods to penetrate beyond the veil of normal perception in an entirely different direction. Shamans, magi, prophets, and high priests used these methods to show their people a sacred vision of the world. This sense of the universe as sacred has existed in virtually every world culture over several millennia. In our contemporary western culture (which is dominating the world through so-called globalization), however, the picture of nature presented by science is often taken to be a severe challenge to the old sacred visions of nature. Since a scientific worldview is the dominant paradigm in our culture and people's religious commitments are among the deepest they make, the issues at stake here are fundamental.

Science and Materialism

Science is restricted to the study of publicly verifiable observations, and hence is restricted to material objects ("material" is used broadly here in the sense of "composed of matter and energy"). There can be no scientific investigation of a spiritual being that acts outside spacetime and doesn't affect the material world. Because science has no interest in non-material entities and because science has been so astoundingly successful in its endeavors, many thinkers have concluded that non-material entities (including any possible spiritual orders of reality) don't exist. Put differently, these thinkers have concluded that the success of science demonstrates the correctness of a materialist ontology. Many commentators have shown convincingly that this argument does not withstand logical scrutiny, but these commentators have been too quick to congratulate themselves for having refuted materialism.

I believe the issues are a little more nuanced than a straightforward logical analysis implies. Along with the purely logical issues, there are also methodological issues. In addition to the actual content of science (what we know and how we know it), there is also the process of doing science (producing new knowledge). A materialist ontology combined with commitment to a scientific analytic process results in a stance that might be called "scientific materialism." It is this view, scientific materialism, that I will often refer to as a *mundane* view of the world, and I will argue that this mundane worldview is in fact the most conducive ontology within which to do scientific work and to understand the results of that work. A mundane apprehension of nature is not logically entailed by the results of science, but it certainly fits very comfortably with those results. The rising fortunes of scientific materialism historically occurring simultaneously with the increasing success of science itself is thus not entirely an accident, nor is it merely the result of aggressive ideologues. Ultimately, I will argue that a mundane view is not only useful but is even, within its limits, correct. In other words, while doing and understanding science, we live in a mundane world. These issues will be examined more carefully in Chapter 3.

Science and Religion

Science, as we've already noted, is only concerned with a limited part of the totally available human experience. Aesthetic judgments, ethical decisions, and religious wisdom all lie outside the purview of science, for example. Religions have traditionally been centrally important parts of most human cultures, and the knowledge claims of religions have both areas of overlap and areas of disjunction with the knowledge claims appropriate to science. Leaving aside many of the social, moral and political roles that religion might play (for good or for ill) in a culture, I am mostly concerned here with the knowledge claims a religion might make. In other words, a certain religious vision of the world might entail some assertions about the way the world is. If these assertions do not overlap with any of the appropriate statements that science can make about the world, then science and religion can neither conflict with nor support each other. For example, a religious assertion is that "God exists" and no scientific result can possibly confirm or deny this assertion. On the other hand, the religious assertion that "the world is 6000 years old" contains a prominent empirical component which can be tested against valid scientific knowledge and found to be wrong. This latter example is a case of conflict, and many prominent writers have promoted the idea that conflict between science and religion has been common and is probably inevitable. Many others (myself included) have noted that this is nonsense and that the problems virtually never turn out to be an "inevitable conflict between science and religion" but instead are simply ill-informed people making invalid claims given their grounds for belief. If proponents of science and of religion are careful to make claims that are truly justified within the limits of their discourse, then most such claims will not overlap and hence have no possibility of conflicting. There are, however, a set of issues and questions in which the overlap is significant and the claims potentially inconsistent; in these cases, Barbour suggests that in addition to conflict and independence there might also be either dialogue or integration between science and religion. A great deal of interesting literature has been devoted to these issues and questions at the boundary between science and religion (see, for example, the works by Russell et al., Barbour, Peters, Artigas, and many others).

The Role of Nature

Where would we expect to find these boundary questions? Since science is a study of the natural world, and since this natural world may well have religious significance, then our conceptualization of nature is likely to be one of these boundary areas. The present work is deeply concerned with the role of nature as a crossing point between the concerns of science and those of religion. Before proceeding further, though, I should clarify what I mean by nature. I don't mean just the beautiful areas of the earth unsullied by civilization, like old growth forests or the arctic tundra. By nature, I mean the natural world *in toto*, including the entire universe and also including human beings as a part of nature. We might speak of the cosmos or of physical reality, and these terms would also be consistent with my broad usage of the word nature in this context. The conceptualization of nature has varied radically in different cultures throughout the ages, and by way of example we examine how several specific cultures have thought about nature in Chapter 2.

How would a religious view of nature differ from a scientific view? There's no single answer to this question, and strictly speaking it need not differ greatly. We've seen, though, that despite the lack of logical compulsion to formulate a single particular scientific view of nature, that the most common way to view nature from a scientific perspective is to adopt the mundane outlook of scientific materialism. Likewise, I would argue that after looking at a variety of religious attitudes toward nature we might conclude that a typical religious view of nature can be best described as *sacred*. The sense in which nature is sacred might very well differ greatly among different religions. After all, religions themselves differ: some have a single deity, some have many deities, some have no deity at all. Nature is crucially and centrally important in some religions, of only minor import in others. Nature is sometimes sanctified by the presence of deity in nature itself, but in other cases by a deity outside nature. In all cases, however, there is some sense in which nature is indeed sacred. We will discuss what this means more precisely and explore the different ways in which nature might be considered sacred in Chapter 4.

Summarizing the view of nature we have developed thus far, we find that nature is mundane (in the sense of scientific materialism) and that nature is sacred (from various sorts of religious perspective). So, nature is mundane, and also nature is sacred. I have identified nature as one of the boundary lines where science and religion intersect, where each perspective has potentially valid claims to make. These two claims (nature is mundane; nature is sacred), however, appear to contradict each other. After further examination in Chapter 3 and Chapter 4, the seeming contradiction only becomes stronger. To make both of these claims constitutes an antinomy, and we might well demand to know which is right and which is wrong. The claims of science and religion conflict, and we need to resolve the problem.

The main thesis of the present work is that both claims are valid and correct. Nature is sacred. I believe that this is true. Nature is mundane. I believe that this is equally true, and furthermore I will argue that there is not necessarily a logical contradiction between these two claims.

Complementarity

In order to show that we can believe both of these apparently contradictory statements (and more importantly that we can live in both of these apparently mutually exclusive worlds), I need to introduce the idea of complementarity. Complementarity is a logical framework for the analysis of ideas in which the exclusive binary relationships between categories found in classical Aristotelian logic no longer hold. Although we will be much concerned with Niels Bohr's famous formulation of complementarity, let's start with a simpler example: Imagine a certain piece of music. Now imagine further that this piece of music is played by a solo violin. How would we describe the music? One option is to specify completely the overtone frequencies in the sound emitted by the violin, which in principle specifies everything we can hear and would thus want to know (it is this kind of information that is engraved on a recorded CD). But this description is solely in terms of the physical sound. Alternatively, we could specify the musical notes, i.e. each pitch and its duration; in this case, there is no sound at all associated with our description, except insofar as the notes can be played (if desired) on some instrument to create a sound. If the instrument is chosen to be a trumpet, the actual sounds will quite different from our first description, and yet the piece of music will be the same. So, we have two alternative descriptions, and they are very different from each other (to the point of being mutually exclusive, at least in a limited sense) since one description is in terms of physical sounds and the other description is in terms of symbolic notations. Of course, the two descriptions are intimately related; they are both describing the same piece of music. Indeed, we can claim that both of these descriptions are necessary in order to have a complete understanding of the piece of music. With only one or the other, we would be missing something essential. This is the essential meaning of complementarity: there are situations in which mutually exclusive alternative descriptions of some phenomenon are not only logically compatible but are both essential in order to have a complete understanding. We refer to such descriptions as complementary descriptions of the phenomenon.

An excellent overview of complementarity, including its relationships to classical logic, to medieval theological thought, and to Kantian philosophy has been given by MacKinnon[1]. More recently, based upon many years of work, Reich[2] has broadened the ideas inherent in complementarity and relabeled the broadened set of ideas as "relational and contextual reasoning" (RCR), which he contrasts with binary, dialectic, and analogical reasoning. Reich carefully examines the logical status of RCR and develops a set of heuristic methods to apply this form of reasoning to issues and problems; a number of specific issues and problems are analyzed in some detail to illustrate the application of RCR thinking and the value it has. An early champion of complementarity was MacKay[3], who has contributed both a rigorous analysis of the logical status of complementarity and who has also urged its use in clarifying the relationships between scientific and religious thought. Mackay emphasizes the status of complementarity as a purely logical concept that defines relationships different in kind from other logical concepts such as contradiction, independence, and identity. It's this added set of possible relationships that make complementarity valuable as a framework for examining concepts, and MacKay notes that this value is especially apparent at the science/religion boundary. "...whether we like it or not, we need it; and by 'we' I mean [...] anyone [...] who wants to avoid logical blunders in seeking to bring science and faith into confrontation."[4] In emphasizing complementarity as a logical relation as opposed to its well-known role in modern physics, however, we need to be careful to not underplay the extremely important work done by Niels Bohr in developing complementarity along several novel lines of thought. It's also important to disentangle the use of complementarity as a generic logical or heuristic analytic tool from the more specialized development of Bohr and the implications of that development.

Bohr's Work

The problem that Bohr set out to solve arose in the context of early attempts to understand quantum physics. The experimental results that were available for understanding the properties of matter and radiation at a microscopic level had become very puzzling. One set of experiments demonstrated unequivocally that both radiation (e.g. light) and matter (e.g. electrons) were wave phenomena: undulations that are not localized in space. Contrariwise, another set of experiments demonstrated unequivocally that both radiation and matter were composed of pointlike particles, quite localized in space and bearing no resemblance to waves. The energies of the particles were found to be in discrete states, which changed discontinuously, and these discontinuities were related in some way to a physical constant of nature, Planck's constant **h**. A more detailed discussion of all this work and its implications is presented in Chapter 6. Eventually, a self-consistent and rigorous mathematical theory was developed for quantum physics, and this theory has been applied to particular physical systems with spectacular success ever since. The major unresolved problem was the interpretation of the theory, i.e. trying to figure out just what the mathematics is telling us about the structure of physical reality. Not surprisingly, the strange features of the experimental results were also found in the mathematical theory. The peculiar discontinuities, the presence of both wavelike and particlelike properties, and the lack of classical determinism were all inherent in the mathematics but not well understood in any sense. In 1927, Niels Bohr presented a piece of work in which he hoped to clarify all these murky points.

The work that Bohr presented was not well understood (or well received) by the physics community, because it turned out that the issues Bohr addressed were not issues in physics but rather issues in epistemology. The heart of his solution to the interpretive problems of quantum theory was complementarity, and Bohr had a specific and precisely defined meaning for complementarity in this context. The starting point for Bohr's line of reasoning is the discontinuous energy changes that occur in the microworld due to the existence of Planck's constant. Bohr refers to this as the "quantum postulate" and it is a contingent fact of nature, not a logical necessity. The quantum postulate is a fact of the world as we find it, and Bohr asserts that this fact carries an implication: due to the quantum postulate, all interactions with physical systems are liable to uncontrollable discontinuous exchanges of energy. The importance of these discontinuities hinges on Bohr's next assertion, which is an epistemological statement. Bohr argues that we only know the properties of a physical system by interacting with it. A totally isolated system has no real meaning for us, because it can disclose no information. Hence, all of our knowledge of the system is acquired through interactions, and all of the interactions include uncontrollable discontinuities. Our knowledge of the world is thus severely limited, not by technological restrictions but by deep issues linked to how we know the world at all. Developing these themes further, Bohr shows that we can define the state of a system if we wish to, but only by giving up any knowledge of the space and time coordinates associated with the system. Alternatively, we can know the space and time coordinates of the system but this then precludes our ability to know its dynamical properties (such as energy and momentum). This latter restriction is extremely important, because conservation of these dynamical quantities is what insures the orderly dynamical behavior of the system (what Bohr refers to as "the claim of causality"). In this way, Bohr arrives at the conclusion that to understand these physical systems we need to use two complementary pictures, that of spacetime coordination and that of cause/effect relations. They are complementary because each picture excludes the other yet both are needed for a complete understanding of the system. Using this framework, Bohr was then able to explain the observed wave/particle duality and the famous Heisenberg Uncertainty Principle in terms of complementarity.

A more detailed and extensive treatment of Bohr's development of complementarity in the interpretation of quantum physics is given in Chapter 7. Excellent expositions of this material are also given by Folse[5] and by Faye[6] (on both of whom I draw extensively in this work) and also by MacKinnon[1]. For our present purposes, we should merely note two further points: First, the combined and integrated spacetime and causal views form the basis of classical determinism, and that's why determinism of this sort is not possible in quantum theory. Second, and perhaps more germane to the arguments we are developing here, Bohr carefully notes that which picture we employ depends on the manner in which we observe the system. In other words, the knowledge we have about a system depends on the experimental arrangement, i.e. on the details of how we acquire this knowledge. This may sound trivial, but it's not; put succinctly, whether we see a wave or a particle depends on how we look. This conclusion of Bohr's development has profound (and controversial) epistemological (and perhaps even ontological) implications. But the implications of Bohr's work, however important, are restricted to the empirical sciences. In order to address the concerns we have raised regarding the validity of both a mundane and a sacred view of nature, we must pass beyond the limitations imposed by such restrictions. How is Bohr's complementarity related to these issues?

Complementarity, Science, and Religion

Complementarity as developed by Bohr and comple-mentarity as a generic logical relations tool have both been employed in the science/religion dialogue, but often in ways that don't properly distinguish between them and sometimes the uses are claimed to be of questionable validity. Criticisms have been offered by Sharpe[7] and by Duce[8] on the grounds that complementarity is an overly limited conception that hinders the attempt to truly engage the two discourses. Other criticism has come from Barbour[9], Alexander[10], and Bedau[11] on the grounds that two complementary descriptions must be of the same logical type, a condition violated in the case of science and theology (or religion more generally). Watts, on the other hand, has presented an extended analysis of the issues that includes consideration of previous criticisms, arguing in the end that science and theology are indeed complementary forms of discourse[12]. Reich[2] also concludes that complementarity (at least in his more broadly rendered RCR formulation) is a highly valuable approach to relating science and theology. Some recent exciting work has explored another variation on these ideas of this sort, known as Generalized Quantum Theory (GQT)[13]. This work draws on the formal structure of quantum mechanics, especially non-commutation, entanglement, and nonlocality. Complementarity plays a central role in this approach to the analysis of issues, and a number of phenomena have been fruitfully explored using GQT. Finally, it's also worth noting that complementarity has been used strictly within theological discourse (by Loder & Neidhardt[14] and by Honner[15], for example) to address issues such as the simultaneous humanity and divinity of Christ.

In many of these cases, both the supporters and the critics of using complementarity to discuss the science/religion relationship face a key issue in whether to use complementarity as developed by Bohr and applied in physics or to use complementarity as a generic logical tool independent of any such use in physics. Much of the inspiration for the application of complementarity in the science/religion area certainly stemmed from the prominence and celebrity it attained in a fundamental part of physical science. Yet all of these commentators, both proponents and critics, point out the major difference between complementarities within physics and the proposed complementarities between scientific and religious discourse. The arguments are over the validity of the application of complementarity in the latter case, and there is sometimes ambiguity over whether a given application is construed as the use of Bohr's version of complementarity or the use of the more general conceptualization going by the same name.

Bohr explicitly intended the complementarity framework he developed to be applied in objective empirical sciences (he had hoped that it would be valuable in many such sciences, especially biology and psychology). Hence, his reasoning must be modified accordingly if we wish to adapt it to problems outside the sciences, as in the science/religion relationship.

Some commentators have expressed the view that complementarity as it is used in physics is so different from what is needed in science/religion discourse that it is essentially useless. A concept forged in the struggle to interpret a physical theory can't be applicable in the broader arena that includes religious faith and claims well outside science. I have already alluded to the limitations built into complementarity as used in quantum physics, but I disagree with the contention that it is irrelevant. There is important content in the careful and sophisticated formulation of complementarity developed by Bohr, content that is not inherent in the more generic version of complementarity as a logical tool. I believe that this content is in fact useful and important for analyzing issues at the boundary between science and religion, and more particularly for the sacred/mundane antinomy that I am exploring in the present work. In order to use the aforementioned content, however, we must follow Bohr's reasoning and suitably generalize this reasoning. Bohr's brilliant contribution to epistemology needs to be generalized and modified at appropriate points to give us a form of complementarity that retains all of his insights but that is not restricted to the empirical sciences and objective knowledge.

Generalized Complementarity

At the heart of Bohr's reasoning is his contention that all of our knowledge about a physical object comes through some interaction with that object. In the case of quantum physics, the interaction takes the form of some experimental apparatus, and this can be precisely specified and described. If we wish to broaden this epistemological lesson to the world at large, what should play the role of the experimental apparatus? In addressing this question, the first thing we notice is that this more general case quite obviously demands the presence of a conscious observer, a knowing subject interacting with the object in the world. (The potential need for conscious knowing observers in quantum theory has been a controversial point of contention; we will assume here, in agreement with Bohr, that conscious observers are unnecessary in quantum theory.) The implications of needing knowing subjects for the acquisition of knowledge have long been a traditional problem in philosophy, a prominent example being the critical philosophy of Kant and his successors. What is new here is to think of this as an extension of the insights gained by examining the problem of knowledge in light of the issues we find in quantum theory, because that is the process by which complementarity takes on a central importance and a new methodological clarity.

Bohr's contentions concerning the inseparability of the observer from the object of knowledge carried a drastic implication: knowledge of the object is no longer independent of the conditions of observation. This aspect of the subject/object relationship is a crucial ingredient in the development of complementarity. Because knowledge of the world depends on the conditions of observation, we need to carefully specify these conditions in order to have any meaningful knowledge at all. In the case of atomic physics, the specifications merely concern experimental arrangements, and the only knowledge we might want is objective knowledge. For the more general subject/object context presently under discussion, knowledge concerning nature will not necessarily be objective. The crucial importance of carefully specifying the conditions of observation, however, is once again mandated by the inclusion of an observer, just as in Bohr's interpretation of quantum mechanics. But our generalization requires that we go well beyond the mere description of an experimental arrangement. For the case of a knowing subject, the kinds of questions being asked; the state of consciousness of the observer; the modes of communication possible and those employed; the role of multiple observers and/or technology used in observation; the effects of culture and history, of time, place, and intention; all these things must be taken into account in order to understand the meaning of any knowledge we may have of nature. Answering these kinds of questions constitutes the methodology that I'm advocating in this work. Performed successfully, a proper specification of the conditions under which knowledge of nature is acquired results in the complementarity framework being free of logical contradictions. The conditions under which nature is found to be sacred are not those under which it is mundane. Both sets of conditions are valid, both are needed for a complete view; and both refer to the same world. This world, I contend, is both sacred and mundane.

The foregoing points are the central message of this work. These points are presented in Chapter 8 in a more detailed form, along with several specific examples of the mundane/ sacred dichotomy (Mount Fuji, the body of a loved one, and a crystal) to illustrate the thinking and methodology. More examples and applications, including extended discussions in some cases, are found in Chapters 10-12. Chapter 8 also includes a discussion of how complementarity relates to monolithic logical integration at one extreme and to dualistic or dialectic approaches at the other extreme. I argue that complementarity is not a form of dualism but rather a special kind of integrated worldview that respects the conditions under which we acquire knowledge. There are still, however, some remaining metaphysical issues that need to be discussed. Does our knowledge of the world affect the world itself? Are these complementary views saying something about reality or merely something about our internal mental constructions? Does objectivity have a privileged status, and if so doesn't this imply an asymmetry between the mundane and sacred apprehensions that privileges the mundane view? These difficult questions are taken up in Chapter 9, where I draw very heavily on the work of Bohr's philosophical mentor, Harald Hoffding. Although I don't pretend to solve these age-old perennial problems of metaphysics, I do argue that no view is truly privileged in a complementarity framework and that if we take complementarity seriously then we are, in the end, making ontological claims about the world.

The Only Sacred Ground

In summary, I claim that nature is mundane (in the sense of scientific materialism) and that nature is sacred (in the sense of living spiritual presence); my claim is not self-contradictory because the mundane and sacred dimensions of nature are complementary aspects of its being. The complementarity framework I use is based on the work of Niels Bohr, but it has been greatly generalized beyond Bohr's formulation because I am considering knowing subjects apprehending nature with no artificial restrictions. We can only have knowledge of the world by interacting with the world, and so our knowledge is only meaningful when we examine the conditions under which we come to know it. This is why the world can be, and is, *both* mundane *and* sacred. We only have one world, but it must reveal itself to us with both of these aspects, and we impoverish ourselves to the extent that we reject one aspect or the other.

Part I

Setting the Stage

2. Ways of Understanding Nature

Before we can explore how nature has been viewed in diverse times and places, we need to know just what the word "nature" refers to in the present context. For example, nature often means that which is not constructed by the artifice of humans; a tree is natural, but a house made from trees is not. Or nature might refer to our own environment on the earth's surface, primarily considering the ecological biosphere but including also the air, soil, rock, and water that sustains life. Or again, there is a long-running debate about whether humans are to be included as a part of nature, as opposed to nature being everything else that is not humanity. In this work, however, nature is intended to be a very inclusive term. All parts of the natural world (living, non-living, human, and created-by-humans) are included, and "world" is interpreted as the entire universe in this context. Put differently, we may equate nature with cosmos (using an older word) or with material reality. The fact that "cosmos" and "material reality" have rather differing philosophical connotations is intentional here; nature might partake of either connotation, and this is precisely the issue we intend to explore.

Concepts of nature have varied radically in different cultures and at different times. To get a sense of the range and richness of such differing apprehensions before we engage in our main project, in this chapter we will look at three of the many possible interesting cases. Forming an appreciation of this variability will be helpful in evaluating the validity of unfamiliar conceptualizations as we proceed later.

Greek Ideas of Nature in Antiquity

The interest of the ancient Greeks in nature, and the way in which they approached nature, varied greatly over the centuries. Not only did their ideas evolve with time, but they also often proposed contesting ideas between the various schools at the same time, with varying degrees of scientific, philosophical, and religious content. The nature concepts of Greek civilization are interesting on their own merits, but they are also highly important due to their influence on later civilizations such as those of the Romans, Islam, and Europe. Not surprisingly, many of the basic underlying presuppositions of modern science have their roots in the thinking of these Hellenic nature philosophers.

Some of the earliest thinking about nature came from the Ionian philosophers. Prior to their work, Greek approaches to nature seem to be mythic and poetic rather than philosophical. In Homer, for example, the forces of nature are personified by the wills of anthropomorphic gods. The Ionians adopted a more rational approach, asking what we could know of the order underlying appearance in the world. For them, the fundamental question to ask turned out to be: what is the universal substance of which all things are made. Thales, the earliest of them, believed that all things originate in water. We may presume that for Thales and his audience water was not the prosaic chemical compound that we think of but rather the life-giving and protean archetypal fluid. But this fluid-ness presents a problem, because we have no clear way to understand how rocks or fire arise from a substratum of water. The ideas of Anaximander address this problem. He proposes that all things are made from a kind of universal principle, an even more protean and unformed substance having no specific properties of its own, which he refers to as the "Boundless" and considers imperishable. Anaximander devised a pictorial model of swirling eddies in the Boundless, separating out propertied substances from which the world is made. An issue with these ideas is the lack of any clear explanation of how these substances that do have properties might arise from the formlessness for the Boundless. Perhaps to address this issue, the third major Ionian thinker, Anaximenes, proposed that all things are made from air. He supplemented this proposition with the idea that air undergoes condensation and rarefaction so as to produce the variety of properties that we see in material substances in the world. This advanced Ionian philosophy by providing a kind of mechanical explanation for how the universal substance could give rise to variety.

These mechanical looking models should not mislead us into thinking that the Ionian philosophers proposed a mechanistic world. The universal substance of Anaximander "envelopes everything, produces everything, governs everything. It is the supreme divinity, possessing a perpetual vitality of its own."[16] Part of the inherent property in the primal substance featured in all of the Ionian systems is some sort of inner directedness toward the ultimate forms it takes. Despite the mechanical rarefactions and condensations of Anaximenes' air, he continues "in thinking of the primitive substance as divine [...] an immanent God identical with the world-creative process itself."[17] And yet, many Ionian explanations of particular phenomena such as lightning and earthquakes are presented as mechanical and pictorial models, as well as important features of their overall cosmological schemas. We see then a key tension in the Ionian conception of nature: the explanations seem to involve what we might think of as causes, but the behaviors are always eventually traced back to inherent properties of the universal substance, which are equated to the divine and have no further explanation. The decisive innovation introduced by Ionian philosophy was the assumption that nature does present an underlying order and regularity that can be explored by the use of reason, and this program was taken as far as it could go based on the limitations of the essential question that the Ionians were asking.

A rather different approach was taken by the Pythagorean school. In addition to their religious and political dimensions, the Pythagoreans developed a nature philosophy based on number. Numbers, and mathematical relationships more generally, are the fundamental basis for the workings of the world. Their thinking is difficult to understand from a modern perspective, because genuine mathematics, number mysticism, speculative reasoning, and observation are all combined indiscriminately into a system composed of parts that seem incompatible yet were synthesized together. One of the key elements of this synthesis was the famous relationship discovered between mathematics and music. All of the consonant musical intervals on the scale are formed by plucking strings whose lengths are integer ratios (2/1 for the octave, 3/2 for the fifth, and so on). This profound connection between number and harmony was central to the Pythagorean work, and it was indeed based on observation of the real world. The investigations of the Pythagoreans into music continued to find broader and deeper relationships to number, and they also performed important work in astronomy (as well as mathematics itself).

There are at least two reasons why Pythagoras and his followers are important. One reason, of course, is the immense influence that the idea of mathematics governing nature has historically enjoyed; we will be returning to this theme often, but the work of Kepler alone is enough to make the point (not to mention virtually all of contemporary physics). The second reason is more directly related to our present story. The Ionian philosophers had identified the key question concerning nature as being about the substance of which it is made, but the emphasis of the Pythagoreans on number shifted the key question away from substance and onto form. The numbers have no substance, yet they are the central concept because they govern the behavior of things. The substanceless numbers are in some sense more real than the substance itself, which merely provides a medium through which the number relationships can manifest themselves. This idea of the superiority of form was to become quite important in Greek thought, and at this very early juncture it provides an alternative formulation to the concept of nature set forth by the Ionians.

Reflection on the issues raised by postulating an underlying universal substance, when the world we observe shows a huge degree of variety, greatly influenced Greek thinking on nature. The question of Becoming, how the things of the world arise from an eternal unchanging Being, was answered in many different ways. Opposing views were proposed by two early giants of Greek philosophy, Heraclitus and Parmenides. For Heraclitus, change was the central feature of reality. All things are process and transformation, not static Being. There is an eternal principle, which he identifies as fire, but this principle itself is the principle of creation, destruction, and transformation. The true underlying order he calls Logos, but it's not clear whether this is within the comprehension of humans. Parmenides, in contrast, maintains that all change is merely illusion and that static Being is the only true reality. The implications of this position are far-reaching: if all that we see is change, and change is an illusion, then everything that we infer from the evidence of the senses is unreliable opinion. The only truth that we can know comes from reason, not the senses. Hence, Parmenides does give us a physical theory of the Ionian kind, but he labels it as merely opinion. These two radical contrasting views were both incapable of leading to any further progress in formulating a concept of nature, but they stimulated thinking by clearly outlining the problem.

This problem of Becoming was addressed by the next generation of philosophers, who devised at least three distinct concepts of nature to solve it. The most direct solution was that of Anaxagoras, who believed that every distinguishable substance possessed its own elemental being. Thus, Being in general was divided into an infinite variety of eternal substances, eliminating the problem of how one substance could transform into another since no transformations occurred, only processes of mixing and replacing. The question now becomes identifying what guides these processes, and Anaxagoras proposed the existence of the Nous (mind, reason, purpose) as the animating principle that acts on the inert and mindless elements to form them into the world we know. In this way, Anaxagoras has now introduced a kind of teleological thinking into Greek nature philosophy, but left many problems by doing so (e.g. is the Nous material or immaterial? in what way does the Nous act? etc.). A much different kind of solution to the problem of Becoming was proposed by Empedocles. He limits the number of uncreated and indestructible elements to four, chosen on the basis of their contrasting properties: earth, water, air, and fire. These four elements are combined together and dissolved apart by the counteracting forces that he terms Love and Strife, which seem to be primarily mechanical in action (attraction and repulsion) despite their poetic names. An important innovation of Empedocles is the idea that any particular material (bone, for example) is characterized by a specific ratio of elements. This introduction of proportions and integer ratios exhibits the influence of Pythagoras. It is this Pythagorean influence that also tempers his tendency to view nature mechanistically. "There is, then, Empedocles concludes, some sort of archetypal intellect identified with Being [...] its thoughts are 'not to be uttered,' but it is the fount from which Love draws the ratios and harmonies for its operations."[18] A third solution to the problem of Becoming, proposed by Leucippus, was to ascribe the unchanging Being to an infinite number of small indivisible bodies, the atoms. All change, then, was due to the movement of the atoms including their coming together into larger objects (or detaching as objects break apart or decay). The atoms exist in a void (which Parmenides had denied existence), and this solved the problem of how motion could occur. Unlike the somewhat similar ideas of Anaxagoras, the motions of the atoms are inherent properties of the atoms themselves and governed only by necessity, i.e. cause and unalterable rules. This rids the system of any hint of duality, but leaves Leucippus with an entirely materialistic concept of nature. We will look more closely at materialism in a later chapter.

Another line of thought accepted the conclusions of Parmenides and set out to defend these conclusions and explore their implications. The most famous member of this school was Zeno of Elea, with his well-known paradox proving that motion is impossible. But this line of thought proves sterile for the understanding of nature, ending in the sophistry of Gorgias (who proves that even Being does not exist). Since there is no point in studying nature, Gorgias and his fellow sophist Protagoras reduce philosophy to rhetoric. Their opponent Socrates is still concerned with a search for truth, but his interests are also in the affairs of humans such as ethics and politics. The concept of nature in this intellectual culture has become unimportant and irrelevant.

Meanwhile, other sources (outside philosophy) of study and information about the natural world were becoming more organized and important. Medicine, for example, was developing its own methodology and conclusions, sometimes independently of those considering nature as a whole and sometimes in concert with them. A good deal of empirical knowledge was acquired (bone and muscle anatomy, for example, and the course of certain diseases), but some of the medical theory (such as the four humors) seems to draw on more general cosmological ideas. The doctors and the philosophers were in agreement that rational necessity (rather than the whims of the gods) governed things, but methodologically the doctors were generally more interested in empirical observation given their agenda of curing the ill. Thus, although medicine did not have the explicit goal of developing a broad understanding of nature, it did indirectly influence the Greek concept of nature through its methods and point of view. Another influence was the development of mathematics, which was making steady progress and slowly becoming more independent of its philosophical roots. This process would later culminate in mathematics as a separate discipline with important applications in astronomy and statics. At about the time that the sophists were beginning to dominate the cultural discourse of Athens, Archytas of Tarentum was making important advances both in mathematics itself and in the Pythagorean program to unite philosophy, mathematics, and nature.

The humanistic concerns of Socrates are combined with the old questions about nature and the cosmos (including the role of mathematics) in the philosophy of Plato. Plato generalizes the Pythagorean idea of mathematical form governing the behavior of matter, and his ideal Forms (or Ideas) include ethical, physical, conceptual, esthetic, and other Forms as well as geometric and numerical Forms. The Forms are eternal and perfect, existing outside of time, space, and matter. Matter itself is formless and chaotic. The world as we know it, i.e. nature, is the result of the perfect Forms impressing themselves on the recalcitrant matter. Plato's concept of nature, then, is that it is an imperfect imitation of the ideal Forms, involving change, growth, decay, and approximation. That which is intelligible in nature is merely a dim reflection of the pure intelligibility of the Forms themselves, and true knowledge can only be knowledge of these Forms. Thus, we can gain true knowledge only through the use of reason; study of the empirical world is useful, but only to gain clues providing grist for reason to work on. In Plato's cosmology, an active principle (identified in the *Timaeus* as the Demiurge or Craftsman) is required to bring order to matter, using the patterns offered by the Forms. There are two crucial elements inherent in Plato's vision of nature: the role of teleology and the role of mathematics. Because nature is a partial realization of perfect Forms, the world is filled with meaning and purpose, and much of the structure and action in the world exists to fulfill these purposes. Because the Forms are often mathematical, the manifestation of the Forms in visible nature gives rise to mathematical regularities there.

Aristotle modified Plato's system in several significant ways. One important difference is that, although he agreed with Plato on the existence and importance of Forms, Aristotle identified the Forms inherently with the actual material things that manifested the Forms rather than some detached immaterial state. From this crucial difference, two corollary differences follow: First, the process of manifesting the Form becomes centrally important, and so teleology becomes one of the major components of Aristotle's concept of nature. Second, the empirical study of the natural world played a much more prominent role in Aristotle's thought than Plato's. The emphases on empiricism and teleology lead in turn to the other major deviation from Plato, namely the considerably less important role of mathematics in Aristotle's thought. Many of these ideas are illustrated by considering how Aristotle would envision the growth of a plant from its seed. Within the seed lies the inherent purpose of becoming the eventual plant, and the growth of the plant is caused by the need for the seed to fulfill this purpose. The natural motions of the stars in circles or of falling rocks toward the ground or of flames leaping upward illustrate again the same way of thinking. Assigning a causal purpose to the qualitatively observed changes in the world constituted an explanation in the philosophy of Aristotle, and this shaped the concept of nature that he developed. Mathematics is not an effective language for explaining qualitative changes, whereas teleological reasoning was well suited for the verbal and logical methods preferred by Aristotle and for the complicated organic phenomena that he empirically observed in such detail.

Although I have focused attention on the approaches to nature developed by Plato and Aristotle as if these were isolated, in both cases we should bear in mind that the nature philosophies were parts of a broader system of thought that included epistemological, ethical, political, ontological, and religious components. These parts are interconnected, and in particular it's somewhat artificial to separate the religious aspect from the concept of nature, since the divine Nous provides the animating purpose and the source of harmony and form in nature. The world is in some sense an organic creature with a mind and soul, and this element of their concept of nature is found in a great deal of the Greek thinking both before and after these two philosophers. We'll return to this point again later. Lastly, in the case of Aristotle, the total collection of work aimed to systematize all knowledge and thought into an overarching synthesis that included nature, humanity, logic, and spirit. Much subsequent work was an elaboration of or a reaction to this corpus.

One strand of development in the Greek view of nature was largely independent of Aristotle's work, however, and this concerned the role of mathematics. Mathematics itself developed considerably, becoming more creative, rigorous, and systematic in the work of Apollonius, Archimedes, and Euclid. The major application of mathematics to the natural world was in astronomy, owing to the high degree of regularity in the celestial motions and large amount of observational information accumulated. Although the Babylonians had employed sophisticated mathematical algorithms to this problem, the great innovation of the Greeks was to devise what we would call mathematical models, i.e. to devise a theoretical concept of how the sun, moon, and planets must move, work out its mathematical implications, and compare these to the actual motions. From the early work of Eudoxus to the culminating system of Ptolemy, this project dominated astronomy, but the attempts of Aristotle's followers to give a physical explanation for these basically mathematical pictures gave rise to issues that would survive the end of antiquity. An aspect of the Greek concept of nature was forming that did not quite cohere effectively.

The work of Archimedes in statics and hydraulics was a rigorous application of mathematics to natural phenomena. Such work was a more modern-looking and less mystical version of the old Pythagorean concept. Although the investigations of Archimedes were brilliant successes, they were too isolated to greatly affect the general Greek concepts of nature (though they did sow the seeds of much later science). Archimedes also performed some practical engineering work, and the practice of engineering and applied science (metal working, agriculture, etc.) in general formed another aspect of the Greek approach to the world in late antiquity. The most famous practitioners in this tradition were Vitruvius and Hero of Alexandria (the latter best known for his small steam-driven device). Yet another practical tradition that continued to make progress in improving knowledge of the world was found in medicine, including the knowledge of human anatomy discovered by dissection. But the knowledge resulting from these kinds of practical pursuits, though important, was not incorporated into any larger vision or theoretical structure. Such knowledge must have influenced the Greek view of nature indirectly, but it remained fragmentary and isolated, not really contributing to a new coherent concept of nature.

This latter project remained the province of philosophy, either that of Plato and Aristotle (and their commentators) or that of three new rival schools that arose: Epicureanism, Stoicism, and Neoplatonism. None of these philosophies, however, regarded understanding the world as their highest priority. Epicureanism and Stoicism were both primarily ethical philosophies, concerned with the right way to live; both devised concepts of nature as a means toward the end of promoting virtue and equanimity in their followers. Neoplatonism was primarily a mystical philosophy with the aim of transcending the material world altogether, and thus did not put much effort into nature philosophy, although some Neoplatonists explored further the idea of mathematics as a more fundamental reality underlying physical nature itself. The Epicureans and the Stoics did put forth substantial views of nature, and these views were almost completely different from each other. Epicurus was a materialist and an atomist in the tradition of Democritus. The major innovation of his system was the introduction of chance ("swerve" in the paths of the atoms) in contrast to the totally deterministic (reason and necessity) system of Democritus. The motivation of Epicurus in adding an element of chance to atomism was to allow free will and therefore moral responsibility to the system. The Stoics, whose views varied somewhat with different teachers and times, adopted the old idea that a void could not exist; all space was filled by their two continuous and intermixing substances, an inert material substrate and an animating vital spirit (pneuma) that gives form in accordance with reason. In the Stoic conception, the universe with its pneuma is a living being, of which the human with its soul is a microcosm. The pneuma is corporeal, and in that sense Stoicism is also materialistic even though forms arise through the action of an Intellect, unlike the Epicurean idea of only blind forces acting.

Nature in Chinese Thought

Although Chinese thinking about nature was not entirely static, it was remarkably stable over a period of two thousand years. In traditional Chinese culture, the elements of philosophy, religion, science, social structure, and political order were highly integrated, so looking at their concept of nature in an isolated way is not sensible. Another complication is the somewhat varying influences of Confucianism, Taoism, and Buddhism on their attitudes toward nature. Yet another complication is the presence of occasional voices and movements that lie outside the main current of cultural thought in China, such as the naturalistic tendency found in the Mohist movement or the skeptical writings of Wang Ch'ung. We can present here only a brief and simplified description of the major trends in Chinese thought concerning nature.

One of the primary cornerstones of Chinese thinking is that it is correlative rather than causal when considering the relationships between events, things, ideas, etc. There are connections between things (north and feminine, for example) that are not either causal or logical relations but rather these connections are said to exist inherently in the make-up of the universe; such correlations are simply a part of the way things are, not the result of anything acting on the things or mutual actions of the things on each other. One major consequence of this correlative thinking is the importance of microcosm/macrocosm interpretations of nature. "A basic feature of systematic thought about the external world as it arose in China is that the body and the state were miniature versions (not just models) of the cosmos."[19] In this view, the human is a microcosm of the entire universe, the macrocosm, and there are a host of correlations between the two. Human society, identified with the state, is also a microcosm of the macrocosmic world. These three realms (human, society, cosmos) are linked together by many particular correspondences and also linked together by the person of the Emperor, who must perform the correct rituals to maintain order in the cosmos and in the society. Beyond its political implications, the microcosm/macrocosm view influenced the understanding of natural phenomena. Various organs and parts of the body are associated with parts of the greater world and their processes and functions operate in the same manner. Just as stagnant water gives rise to stench and decay in a pool, so stagnation of the body's vital fluids leads to illness in a person. The emphasis is much more on process than on structure (anatomy in the body, physical character-istics in the world). "...organs and tissues figured in medical doctrines as mere correlates of the body's systems of functions, mainly useful in diagnosis and in schemata that aligned parts of the body with physical features of the macrocosm. [...] circulation is fundamental not only to the body's growth but to its maintenance, irregularities in it are responsible for pain and disease. Somatic blockages are analogous to failures of circulation in the universe and the state."[20] In astronomy, the order observed in the celestial realm exemplified the desired order in a properly functioning Confucian society/state, and the particular sky events as recorded or predicted formed the basis for the calendar, presented by the Emperor as a mandate for the existence of a new year. Clearly, the Chinese interpreta-tions of astronomical phenomena were heavily influenced by the microcosm/macrocosm view.

There remains a question of what governs all these correlated events and ideas. Once again, the answer involves inherent properties and relations as opposed to external causal influences. More specifically, the Chinese conception of nature is an organic one, in which processes and events unfold in accord with their inherent tendencies, and relationships are maintained by preexisting harmonies. "...the *philosophia parennis* of China was an organic materialism. You can illustrate this from the pronouncements of philosophers and scientific thinkers in every epoch. Metaphysical idealism was never dominant in China, nor did the mechanical view of the world exist in Chinese thought. The organicist conception in which every phenomenon was connected with every other according to a hierarchical order was universal among Chinese thinkers."[21] No central directing intelligence is required or found in this paradigm. Changes and processes simply occur as they are meant to, in conformance with their own inner natures and in relationship to all other changes and processes occurring similarly. Immaterial agents and entities are not a part of orthodox Chinese philosophy, but this fact must be qualified in two ways: First, a strain of older animistic thinking did survive in Chinese culture and sometimes became integrated into the orthodoxy. For example, to say that the inner nature of water is to flow downhill can be closely related to (or even shade into) the attribution of a "desire" to flow downhill on the part of the water or the spirit that animates and governs the behavior of the water. The second qualification is that the substances of which all things are made, although material, have a surprising and peculiar set of properties compared to material substances found in many other cultures. These properties will be more apparent soon, when we examine more closely the details of the system that evolved, but an important point to reiterate is that Chinese thought was much more concerned with process and function than with substance *per se*. This emphasis governed the questions that were asked and the conceptualization of "material substance" that ultimately was formulated.

The basic substance underlying existence in Chinese thought is called *ch'i*, but to simply call it a substance is rather misleading. The word does not have any simple equivalent in English, though it is sometimes rendered as "vital principle" or "energy" and so on. "The most basic stuff that makes the cosmos is neither solely spiritual nor material but both. It is a vital force. This vital force must not be conceived of either as disembodied spirit or as pure matter."[22] So although *ch'i* is a kind of substance, it transcends dualities of matter and spirit, of active and passive, of structure and function. All of these potentialities are inherent in the concept of *ch'i*, but in order to develop them further we need a more extensive set of ideas. The undifferentiated unity allows no further discussion; the primary bifurcation of this unity is into the great archetypal polarity, represented in Chinese thought by *yin* and *yang*. Each has a large set of correspondences (earth, feminine, north, moist, cool, etc.; sun, masculine, south, dry, hot, etc.) and these can be used to describe the state and action of the *ch'i* as it circulates, vivifies, and imparts form to its own self in the continuous foundation of the universe and the human being. Once again, it is movement, process, and function that dominate this analysis, rather than static structures or properties of substances. In addition to the duality of *yin* and *yang*, a further set of descriptive categories is provided by the *wu hsing*, the five phases (sometimes translated as five elements, five agents, and so on). The five phases (earth, fire, water, wood, and metal) are once again not merely static substances, but rather they are modalities of transformation for the *ch'i* itself, and once again each of the five has associated with it a long set of correspondences. Together, the concepts of *ch'i*, *yin* and *yang*, and the *wu hsing* comprised a powerful and flexible system with which to describe the universe, the body, and the state, along with their correspondences and interrelationships. "A fully developed cosmological doctrine, in which *yin-yang* and the five phases became categories of *ch'i*, tools for analyzing its complex configurations and processes, appeared in the first century B.C."[23] From its origins in the Han dynasty, this system lasted almost two millennia and was applied in medicine, political theory, alchemy, astronomy, and ethics. Indeed, it is still used in traditional Chinese medical practice.

This complex set of ideas served to answer questions that arose within the context of China's organicist paradigm. Given that actions occurred spontaneously in accordance with the harmonious unfolding of their inner natures, we want to know how to explicate these inner natures to understand the world (or behave correctly), and this system is designed to offer such understanding. The other great concept in Chinese thought, which underlies all the rest that we've discussed, is the *Tao*. The *Tao*, or Way, cannot be described or explicated itself, but it is the metaphysical foundation for all of the processes and functions encompassed by the *ch'i*, *yin* and *yang*, and the *wu hsing*. Lastly note that all of the myriad correspondences implied by the microcosm/macrocosm picture can be described, worked out, and categorized within this system. But the correspondences are not correspondences of static structure; the Chinese concept of nature (which includes the human and society as well as the cosmos) deeply emphasizes dynamic process at all levels. Their concept of time and change, though prominently featuring cyclic thinking (such as the changing of the seasons) and not having anything like time as a linear dynamical variable, does clearly offer a sense of time as having reality and directionality, a sense that novelty occurs even as an ideal of stability is sought.

The correlative basis and great complexity of this system hide two other important aspects of the Chinese attitude toward nature. One aspect is the strongly empirical side of their thinking. In astronomy, for example, precise observations were recorded for thousands of years. In medicine, a large amount of empirical information about symptoms and treatments was accumulated and passed on, also being folded into the system described above. The alchemists certainly included an empirical element in their work, and the state sponsored data collection on rainfall, seismic activity, etc. But another important aspect of Chinese attitudes toward nature stems from the mystical Taoist influence on their culture, namely the receptivity, openness, and love of nature that lie underneath the many complex correspondences, and an intuitive grasp of natural processes derived ultimately from union with nature, decreasing the separation to nothing. Although mixed with many other influences, this attitude remains a part of the Chinese concept of nature through the years.

These multiple influences give the Chinese concept of nature a deep richness and subtlety. The Confucian concern with the correct functioning of society and the role of the state reinforced the idea of a microcosm/macrocosm relationship, for example. And the absolute power of the Emperor deployed by a complex bureaucracy, to which almost all intellectuals belonged, shaped the idea of correlated but uncaused relationships. Also, the membership of the intellectuals in a bureaucracy that owed all of its power to the Emperor tended to maintain stability in the ideas about nature, as did the cultural trait of imputing authority to ancient sages as opposed to recent innovations. Interestingly, many practical workers, such as engineers, shipbuilders, metalworkers, and rural medical practitioners were barred from membership in this bureaucracy, so these highly skilled people were generally not literate and their knowledge had little effect on the theoretical ideas of the more literarily inclined state-sponsored intellectuals (alchemists also fell into this category). Hence, a great deal of the first-hand knowledge about the workings of nature in China was not well integrated into the rest of its intellectual culture, and yet was clearly present in the culture overall.

While some areas, like dynamics and anatomy, were not much emphasized, a number of particular disciplines were highly developed in China. Astronomy, for example, engaged in an unbroken series of precisely recorded observations that lasted millennia. A sophisticated coordinate system was used to record this data, and constantly improved mathematical algorithms were devised in order to make predictions based on the regularity of the night sky. This mathematics, however, was not geometrical (but instead arithmetic/algebraic), and the Chinese did not attempt to create any kind of model for the motions of celestial bodies. Such a project would have been antithetical to their entire view of the world; the order and regularity of these sky events simply arose from the broader order of the entire cosmos, micro and macro alike. That this order could be described with mathematics was interesting but not extraordinary, because order was naturally to be expected everywhere in accordance with the *Tao* and the organicist conception of nature. From it came the accompanying order to the calendar, presented each year by the Emperor to the people as a sign of his maintaining of this proper cosmic balance in the state and in nature itself. To ask for some sort of other underlying source of the order and regularity found in astronomical observations would have made no sense, because the observed order was simply a manifestation of what was inherently there in all of nature, even when disguised by locally unbalanced tendencies (this is why an unexpected event was considered a bad omen for societal affairs).

Chinese cosmology has already been described in some detail. The major elements of cosmology were the macrocosm/microcosm paradigm, the material organicist conception of the world, the role of *ch'i, yin* and *yang*, and the *wu hsing*, and underlying all of these the ineffable *Tao*. Like any genuine cosmology, in the traditional sense of the word, this conception penetrates into the specific understandings developed in all of the particular sciences. This is seen in astronomy, and we have already discussed the influence of the Chinese cosmology on medicine. These cosmological ideas were combined with a store of empirical knowledge, resulting in a sophisticated medical practice that developed its own diagnostic methods based on observable symptoms and its own treatments such as acupuncture, moxibustion, and herbalism. One of the basic ideas of this medicine was that of disease as a disruption in the natural balance and harmony of the body (along with its relation to the world). Balance in the *yin* and *yang* plus proper flow of the *ch'i* through the body are essential to good health, so problems with these things as demonstrated by observed symptoms were treated accordingly (prescribing a *yang* herbal regimen to remedy a *yang* deficiency, for example).

The herbal remedies were supplemented by mineral-based drugs supplied by the alchemists. One of the major overarching projects of Chinese alchemy was to prolong human life, producing the long-sought elixir (this goal eventually became incorporated into European alchemy by way of Arab alchemy). Although Chinese alchemists are sometimes represented as superstitious and uncritical practitioners of a degenerated form of *Taoism*, they actually produced a good deal of genuine chemical knowledge (e.g. ammonium chloride, ammonium carbonate, and potassium nitrate) and developed instrumentation and techniques that also influenced the Arabs and hence the west. The theoretical ideas that guided the alchemists, though much different from the understanding of modern chemistry, were complex and sophisticated, related to the overall cosmology of China but more independent of it than the sciences supported by the state bureaucracy, retaining more of the archaic elements from prehistorical culture. At an empirical level, though, quantitative methods were employed and recorded. Chinese alchemy, though never incorporated into the mainstream of Chinese culture, was an influential and successful enterprise, affecting medicine, metallurgy, and even (as we shall see) warfare.

Mathematics became highly developed, but only in certain areas such as algebra and arithmetic. Deductive systematization did not become important, with the emphasis remaining on practical calculational methods (especially simplified algorithms that could be implemented by relatively unskilled members of the bureaucracy; the state collected and used large amounts of quantitative information). Geometry in general was not developed to any large extent until after contact with the west *via* the Jesuits. Although not having geometry did not hinder work in astronomy or engineering, it may have inhibited the study of optics to some extent. The role of number in Chinese culture was important and had a quasi-mystical significance, as shown by the importance attached to the four directions, the five phases, the nine heavens (as well as, mathematically, the nine numbers of a magic square), and the sixty-four hexagrams of the I Ching. This aspect of number also reinforced the correlative thinking of Chinese cosmology and thereby permeated the thinking of all intellectual culture. The extent of the mutual influence of numerical significance and practical mathematical computation is not clear, but both certainly had a role beyond mathematics itself.

Engineering, although important and supported by the state, was not part of the official Imperial bureaucracy. For this reason, engineers were not literate and have thus left few records of their thinking. Their accomplishments, however, in areas such as massive hydraulic and irrigation projects, weapons technology, the iron chain suspension bridge, and shipbuilding were highly impressive. Cloth and paper making, metal work (including iron casting many centuries before Europe), temperature control in the porcelain furnaces, and a host of other accomplishments attest to the practical technological skill of the Chinese. We do not have a clear understanding of how this body of technological knowledge was related to the cosmological conception of nature developed within the literate elite, but it's fair to speculate that both were fed by the empirical strain of Chinese thought and by the strong appeal to tradition.

We'll finish by looking at a few of the most famous results and accomplishments of the Chinese. The invention of the compass is certainly prominent among these accomplishments, and this invention was the result of their study of magnetism. Magnetic phenomena were less mysterious in the context of China's idea of nature than in that of Europe, because the non-causal relationship of things separated by distances was already a part of Chinese thinking. That a bar magnet should orient itself in a certain direction, subject to no apparent push, was a natural outcome of their organicist and correlative thinking. They also studied this phenomenon empirically with great thoroughness, working out the details of declination at an early date. That the Chinese soon put their knowledge of magnetic properties to practical use in the compass is not surprising given their usual practical attitude. A different accomplishment for which they are known is the development of acupuncture. This technique is now becoming more often used in the west, and controversies over its efficacy and mechanism in particular cases don't detract from its status as a major success of the Chinese medical tradition. Once again, we see the combination of Chinese cosmological theory and a great deal of empirical observation as leading to the results under discussion. Yet another major accomplishment of the Chinese was the great astronomical clock tower of Su Sung, built in 1088 at least three centuries before any similar clock in Europe, and which "was preceded by the elaboration of a special theoretical treatise by his assistant, Han Kung-Lien, in which the trains of gears and general mechanics were worked out from first principles. He did not have Euclid, but he could do that."[24] Here, the integration of cosmological conceptions and practical know-how is seen in the purpose of the device as well as in its construction. Finally, we consider the perhaps most famous Chinese invention of all, gunpowder. Gunpowder eventually resulted from the studies of the alchemists, who mixed saltpeter (potassium nitrate) together with sulfur for various purposes as early as the 7th century, as recorded in alchemical treatises from that time. They noticed that these mixtures burned violently, and the later additions of carbon sources into the mixture enhanced the effect. Reports of such combinations appear by the middle of the 9th century, and by the end of the 10th century gunpowder as we know it has been given a name (*huo yao*) and is being used in warfare to make simple bombs. The further evolution of gunpowder's use in military technology doesn't concern us here. What is interesting about this story is the advanced and advancing state of alchemical knowledge that it suggests at these early dates, a suggestion that can be generalized to the sophistication of the entire Chinese encounter with nature.

Nature Concepts in 12th Century Islamic Civilization

Prior to the time of Muhammad, Arab tribesmen had already developed a stock of knowledge concerning nature in order to survive in the harsh desert environment. Knowledge of plants and animals was obviously useful, for example, and knowledge about the night sky was needed to navigate. The world-view of these Arabs had an element of animism, so nature was also a living presence to them. The Islamic revelation, embodied in the Quran, did not add any specific pieces of information regarding nature, but it profoundly reoriented the entire approach to nature experienced by a believer. The Quran "provided the earliest stimulus for reflection on nature. The Quran contained a large number of verses that called attention to the harmony, symmetry, and order present in the natural world [...] This invitation to reflect on nature was such an insistent theme of the Quran that no one could ignore it..."[25] Into this intellectual milieu stimulated by the Islamic revelation, a huge amount of new information and novel traditions of thought were soon introduced by the rapid conquest of many ancient and learned cultures. The learning, including the approach to nature, of the Greeks, Egyptians, Babylonians, Persians, and Hindus were all added to the growing intellectual tradition of Islamic civilization. This legacy was reworked and synthesized in the light of the Islamic revelation (though not without conflict and controversy), and used as the basis for a great deal of original work by Muslim thinkers. The result of this process was the Islamic concept of nature that emerged during the early Middle Period, from about the 11th to the 13th centuries.

Among the early intellectual traditions of Islam was a method of theological inquiry that came to be known as *kalam*. The basis of *kalam* was an absolute belief in the Quranic revelation, but the Quran leaves a number of important questions unanswered. Issues such as how to reconcile predestination and free will, or whether God's total justness is consistent with both omnipotence and the presence of evil in the world, were growing in importance. Although the Quran served as the starting point and fundamental touchstone for grappling with these issues, a rational analysis was also needed to deal with them fully. Once these sorts of discussions started, several different schools arose proposing different views. Although the initial discussions were centered on purely theological issues, it became inevitable for the conversation to expand into cosmological doctrines as well, and in this way *kalam* disputation entered into the development of the Islamic concept of nature.

Meanwhile, an entirely different tradition was also developing, based on the entry of Greek philosophy into Islamic culture. An extensive translation movement from the 8th to the 10th centuries brought many Greek philosophical texts into the Arabic speaking world, even as Arabic itself evolved into a more precise, flexible, and powerful instrument for working with ideas. These Greek works on metaphysics, logic, mathematics, and natural sciences inspired a new approach to learning within Islamic culture, the tradition known as *falsafah*.

The adepts of this tradition, the *faylasufs*, were dedicated to a life of reason and critical inquiry. Their view of nature emphasized the orderly workings and unchanging principles of the universe, and hence the *faylasufs* became important builders of the Islamic scientific tradition. The emphasis on rationality and inherent order in *falsafah*, however, were not easily reconciled with the fundamental revelatory basis of Islam. For many Muslims, challenging the authority of Muhammad and the Quran by putting the authority of rational thought uppermost was offensive. "The Socratic tradition could not rest content with being bound to limit its questioning within a framework which was imposed by a historical intervention such as Islam. Nor could the Quranic tradition accept subordination for its conclusions to the authority of private human speculation."[26] After a considerable amount of controversy and accommodation, taking place over several centuries, a modified *falsafah* developed in ways that were more integrally related to the evolving Islamic culture. Although certain segments of society continued to mistrust and resent *falsafah*, it had become an important part of the mainstream of Islamic civilization by the 12th century, and a crucially important contributor to their concept of nature. Within the overall tradition of *falsafah* were embedded a number of particular sciences (astronomy, mathematics, medicine, optics, and alchemy) that flourished for several centuries, the most advanced in the world of that time.

The role of nature within Islam was somewhat problematic. The central message of Muhammad concerned humans, and how they should relate to God and to each other. Hence, ethical study and teachings along with interpretation of the Shariah law (culminating in a complex system of jurisprudence known as *fiqh*) became paramount. Furthermore, because Islam is profoundly historical, in the sense that the revelation to Muhammad was a specific historical event though it has cosmic significance, the study of history is also considered of great importance. Only the Quran and the hadith (sayings of the Prophet) were irrevocably worthwhile in Islam, and for some Muslims nothing else had any worth at all. And yet, nature did have a valid place in Islamic culture, based on both the Quran and the hadith themselves. "Muslim religious doctrine promotes a concept of the entire material universe as a sign of God's activity, [...] Thus, in order to understand God, it is necessary to investigate every aspect of his creation—all phenomena that exist in the world [...] study of [God's] activity is thought to provide knowledge of the right path toward the proper life...."[27] The study of nature also offered practical benefits, in areas such as medicine for example. These motivations, along with innate curiosity, assured the development of nature concepts within Islam, even if overshadowed by the study of *fiqh* law, Arabic grammar, and so on. Another factor that hindered the study of nature in Islamic culture, however, was the educational system. Areas like astronomy, mathematics, and medicine were not taught in the organized system of *madrasa* schools; these advanced fields could only be learned by apprenticeship with established scholars at observatories, hospitals, royal courts, and settings of that sort. Despite all this, the study of nature reached heights in the Islamic world that Europe would not see for half a millennium. More importantly, the concept of nature they ultimately developed was fully integrated into the fabric of their entire culture.

In the sophisticated cultural milieu of the 12th century, all forms of knowledge and behavior, including the understanding of nature, were grounded in the fundamental basis of Islamic life, the Quranic revelation. Nature was accordingly understood as a manifestation of Divine will, and any particular study or discipline was interpreted within this context. "Islam and science discourse existed within the larger intellectual tradition of Islam and although there were many foreign currents that ran through the warp and weft of the tradition, it remained integrally linked to the Islamic worldview [...] science in the Islamic civilization was part of a larger tradition of learning that arranged different disciplines in a hierarchical structure like the branches of a tree. The trunk of the tree in this case was none other than the central concept of Islam: the Oneness of God (*Tawhid*). Because of this central unifying concept, all branches of knowledge, including the natural sciences, were linked through an inalienable nexus with the metaphysical concepts of Islam. Each branch of knowledge was a contributing tributary to the main stream."[28]   An important characteristic of this hierarchy of knowledge was that the physical and the metaphysical were organically related within it, not compartmentalized separately in the way that the modern world typically thinks. Empirical studies, metaphysical insights, mathematical models, and religious experiences all blended together in their emerging concept of nature.

This intrinsic connectedness is seen most clearly in the centerpiece of the Islamic nature concept, cosmology. The Quran itself contains a number of verses outlining cosmological processes, and these verses formed the basis for earliest *kalam* treatment of cosmological questions. The contact with Greek thought introduced the idea of an eternally existing universe, as opposed to a universe that was created at a specific beginning of time. The fundamental conflict between these two formulations drove a great deal of cosmological thinking in the *falsafah* traditions. These tensions were ultimately resolved, in part by invoking the emanationist ideas of Plotinus, and an Islamicized version of Hellenic thinking evolved, though the subject always remained one of ongoing debate. The epitome of this work is found in the thinking of the great scholar Ibn Sina (Latinized as Avicenna), who successfully synthesized Aristotle with the Quuranic revelation. Ibn Sina emphasized the distinction between essence and existence, so that these two attributes were both found only in the Necessary Being (that is eternal) but not in material reality (that is merely contingent). "But Ibn Sina's concept of 'Necessary Being' is used here as an ontological principle, in the context of a cosmology where modalities of necessity and contingency play a crucial role. However, it is the close affinity of Ibn Sina's Necessary Being to the Quranic God and of his single order of reality to the general thrust of Quranic teachings that makes this Hellenized scheme a compelling and powerful case of a fundamental recasting of the Greek legacy [...] In sum, Ibn Sina's cosmology rests on two fundamental premises: (i) that a material entity can emanate from an intelligence, and (ii) that there is in some sense a unity in the entire universe which is dependent upon the fact that the Necessary Being is the ultimate cause of every entity."[29] Islamic cosmology, however, like other traditional cosmologies, was not concerned only with the origin of the universe but also with the purpose of the universe and with the role of humans within this purpose. Humans constitute a microcosm of the greater cosmos, and serve as a link between the spiritual and material realms of existence. Cosmology, with its metaphysical and religious overtones joined to more physical ideas, served as one of the most important links between the concept of nature and the central concerns of Islam.

Astronomy was one of the most important of the Islamic sciences, relating genuine observations of nature with mathematical calculations and models yet also making contact with the more speculative cosmological ideas. Knowledge of the night sky had a long history in Arab culture, was explicitly referred to in the Quran, and was the subject of the earliest translations of Greek and Persian texts. Ptolemy's *Almagest* was quickly incorporated into Islamic astronomy, and its observational data continuously improved. The Muslim astronomers were excellent at measurement and observation, establishing major observatories at Isfahan, Maragha, and Samarkand plus a large number of smaller installations (Rayy, Shiraz, Baghdad,

Rakka, Cairo, Ghazna, etc.). These observatories were outfitted with improved instruments of various sorts, such as large measuring arcs, sophisticated versions of the sundial, astrolabes, quadrants, and celestial globes. The astrolabe, in particular, was perfected to a high degree. Increasingly precise measurements were compiled into extensive astronomical tables (known as *zij*) and were used for both practical and theoretical purposes. Practical uses included improvements of the calendar, timekeeping, and establishing the direction toward Mecca (all needed for purposes of Islamic ritual and prayer). Theoretical issues involved the improvement of Ptolemy's complex model of the planetary movements, which employed epicycles, deferents, and equants. Better data induced Muslim astronomers to look critically at the assumptions and methods of the Ptolemaic system, resulting in a literature that extended over several centuries. A particularly pressing question involved the lack of uniform circular motion resulting from the use of equants. This problem was solved by Nasir al-Din al-Tusi, who developed a mathematical construction that preserved uniform circular motion, now popularly referred to as the Tusi Couple (there is evidence that Copernicus was influenced by Ibn al-Shatir's use of the Tusi Couple in astronomy). The *zij* tables were also useful in astrology, further connecting the order in the night sky to order at all levels of the cosmos. Islam produced many thinkers who made important contributions to astronomy, among the most noted being al-Biruni, Ibn Qurra, al-Battani, al-Khwarizmi, al-Tusi, al-Haytham, al-Shirazi, al-Urdi, Ibn Bitruji, al-Juzjani, and al-Shatir. The relationships between astronomy and mathematical work, Aristotelian physical thinking, and the metaphysical conception of order in the cosmos made astronomy a central element in the concept of nature in Islamic thought.

Mathematics itself was one of the most sophisticated parts of the Islamic intellectual tradition. Ideas of logical structure and geometry were inherited from the Greeks, advanced computational methods from the Babylonians, and advanced number theories came from India. The idea of zero, and its use as a place-holder in a decimal number system, was adapted from Hindu thought, improved upon, and ultimately transmitted to European culture as the famous "Arabic numerals" that we still use today. Important advances in working out the use of this number system to make computations simply and efficiently (i.e. arithmetic) were pioneered by Muhammad al-Khwarizmi, one of the greatest mathematicians in history. Al-Khwarizmi is also famous for another major achievement, namely the development (in some ways, the invention) of algebra. Both "algebra" and "algorithm" are words based on terms coined by al-Khwarizmi. Islamic mathematicians like al-Haytham, al-Tusi, and Omar Khayyam also made important advances in geometry and trigonometry. "It was only much later, in the nineteenth century, that the medieval Muslims were understood to have defined what came to be called Euclidean geometry, and that, without realizing it, they had pointed the way toward the discovery of independent non-Euclidean disciplines."[30] Algebra continued to advance, with a variety of solutions found for second and third degree equations, and Khayyam also systematized these advances. The mathematics of conic sections, inherited from Alexandrian Greek culture, was also studied and advanced by Muslim mathematicians. Mathematics, along with its intrinsic importance, played roles that were both physical and metaphysical in Islamic culture. Mathematics was applied to understanding musical acoustics and optics, as well as astronomy. But in addition, the Neoplatonic and Pythagorean understanding of mathematics as the mystical substratum of reality was also deeply embedded in Islamic culture and related to the cosmic order revealed in the Quran.

In the study of music, especially, the physical and

metaphysical aspects of mathematics blend together. Muslim studies of music continued along the lines begun by the Pythagoreans, relating musical intervals to mathematical integer ratios. Studies of this sort were written by al-Kindi, al-Farabi, and the Ikhwan al-Safa (Breathren of Purity), unifying musical intervals with cosmological and astronomical ideas by means of mathematical ratios. Meanwhile, another area in which mathematics played a prominent role was the study of light, optics. The study of optics was arguably one of the premier Muslim contributions to science. Al-Kindi studied the reflection of light and visual perception, while al-Razi and Ibn Sina wrote on optics from an Aristotelian perspective. The towering figure of Islamic optical investigations, however, is Ibn al-Haytham (Latinized as Alhazen). Al-Haytham performed experimental studies and mathematical analyses of refraction and reflection of light, formation of rainbows, parabolic mirrors, and visual perception. His theory of visual perception was close the modern idea, involving image formation by the eye of light emanating from the perceived object. Al-Haytham's work in the areas of rainbow formation and the camera obscura were later continued by al-Farisi, who considerably improved the explanation of rainbows (which had also interested the Ikhwan al-Safa). The work of al-Haytham also influenced the thinking of European figures such as da Vinci and Roger Bacon.

Medicine was another discipline for which many treatises were translated almost from the time of the Prophet. Gondeshapur, in Persia, was already an important center of medical information and practice when the Arabs conquered it in 638. The Greek medical heritage of Hippocrates, Galen, and Dioscorides became available, along with knowledge of Indian medicine and the practical skills of the Nestorian community. In addition, the wealthy and well-organized Caliphate established large hospitals where medical expertise was concentrated, improved, and transmitted to students.

Many prominent *faylasufs* were actually primarily physicians. Two of the prominent early examples of this category are Ibn Sina and al-Razi (Latinized as Rhazes). They both wrote encyclopedic medical reference texts, and Al-Razi is also noted for his work on smallpox. Muslim physicians continued to improve on the legacy they had inherited from Greek and Hindu sources, investigating human anatomy, contagious diseases, and pharmacology, eventually establishing a literature critical of the old sources and writing new medical texts based on their own discoveries. Among the more prominent figures in this movement were Ibn Rushd, Ibn al-Nafis, Ibn al-Khatib, al-Zahrawi, and Ibn Zuhr; among their accomplishments were studies of blood circulation, the anatomy of the eye, the idea of contagion, and the development of advanced surgical techniques and instruments. Also worthy of note is the vast pharmacological compendium written by Ibn al-Baytar, with more than a thousand medicinal plants included. In addition to the simple herbal remedies, Muslim pharmacology also began to include prepared drugs based on alchemical processes.

Alchemy was an important, if sometimes esoteric, part of the Islamic intellectual tradition, and the application of alchemy to pharmacology was only a small part of the entire field. The Muslims inherited the strong tradition of Alexandrian alchemy from Hellenic culture and later added to this the alchemical ideas if India and China. The ancient sources of alchemy in craft traditions, metal smelting, and shamanic practices had already been transformed by contact with both philosophical and mystical currents by the time it entered Islam. All these aspects of alchemy, both practical and esoteric, were retained and amplified within Islamic culture, and new elements of experimentation and quantitative thinking were added. The most important early figure by far in Islamic alchemy was Jabir ibn Hayyan (Latinized as Geber), who lived in 8th century Baghdad and to whom many treatises are attributed. The central idea of Jabir's work is "balance," a concept that seems to refer simultaneously to numerical amounts of differing elemental principles of substances and to spiritual and physical attributes of these substances. Some of the later alchemical investigators, such as al-Biruni, al-Khazini, and al-Razi emphasized more the quantitative aspects of the balance idea. Other alchemists, however, emphasized the more esoteric aspects of the work, in which the changes of material substances directly reflect changes in soul of alchemist himself. Alchemy in this sense is directly related to the cosmological doctrines of humans as a microcosm of the universe, and Jabir's work is related to the Quranic verses dealing with the balance. In contrast, al-Razi "demonstrated a firm preference for proof through experiment [...] basic alchemical processes such as distillation, calcination, crystallization, evaporation, and filtration gained precision [...] the standard alembics, beakers, flasks, funnels, and furnaces began to resemble those of modern times."[31] Islamic alchemy cannot be simply reduced just to chemistry, or to magic, or to psychology, or to philosophy, or to spiritual purification, or to any single aspect; all of these elements operate together in alchemy, which in many ways makes it a paradigmatic example of the Islamic concept of nature.

It's difficult to make generalizations about a culture that spanned many centuries, included lands from Spain to India, had many different social classes, and ideologically competing elites (the pious *ulama* scholars, the *faylasufs*, the courtly literary *adibs*, the Sufi mystics, and so on). "...the Shariah-minded guardians of the single godly moralistic community maintained a frustrated tension with the sophisticated culture of Islamdom, which they could successfully condemn but not effectively destroy."[32] And yet, the *faylasufs* did manage to forge a synthetic concept of nature that was characteristically Islamic and highly successful, even if not the centerpiece of the culture. The centerpiece of the culture, of course, was the Quranic revelation, which (along with the Arabic language) held the civilization together. The activities of *falsafah* needed to accommodate and ultimately merge with these central aspects of Islam, but *falsafah* also provided numerous practical services in support of them. The faithful needed to know precisely the times of the day for prayer and the direction of the Ka'ba from any geographical point, information provided by the astronomers and mathematicians. Mathematicians also supplied techniques for the division of inheritances in accordance with the dictates of the Quran, and the times for religious observances in the lunar calendar. But beyond these practical considerations, the *faylasufs* offered an understanding of nature integrated with the overall Islamic vision of reality. It was not a static or simple understanding, as evidenced by al-Ghazali's critical response to Ibn Sina in order to revitalize the spiritual roots of Islam, but the fundamental idea remained consistent.

This fundamental idea was a concept of nature which was not separate from the concepts of ethical behavior, spiritual reality, historical sense, or even legal rules. All of these aspects of Islamic culture were combined into a hierarchical system stemming from the Quranic revelation. "...they all sought to explain the cosmos in the light of revelation, in particular, in the light of the doctrine of *al-Tawhid*, the Unicity of God, which made it impossible for two cosmic orders to co-exist. This fundamental principle acted as a prism through which all theories were passed in order to test their validity. It was this powerful doctrine, situated at the very heart of the Quranic message, that made it possible for the Muslim scientists to transform those Greek theories about nature which conflicted with revelation [...] It was through the inherent power, simplicity, and uniformity of this principle that was operative in all realms of knowledge that a coherent Islamic worldview appeared."[33] Nature did not exist as a separate thing apart, but instead nature was like all else a manifestation of the divine will such that all parts of nature had a proper place and meaning within the overall order of reality.

3. The Mundane World

Brief Historical Introduction

Materialism is a view of nature that assumes nothing is real except matter and energy behaving according to set rules. Although there are subtle variations among various materialist philosophical schools, all of them share certain core assumptions, such as: the lack of any possible immaterial substances or agencies; the absence of any beginning or end to the material making up the universe; and the purely material basis of mind and consciousness. "This plenitude of being has no gaps, breaks, or noncontinuities. There is no immaterial or supernatural region zoned off from it [...] Terrestrial life is an accidental realization of one [kind] of being possible to material substance [...] Human thought and feeling [...] consists of neural events that individually are as insensitive, unthinking, and unfeeling as all other basic chemical reactions...."[34]. Since the posited material is governed by rules and since it is the job of science to ascertain what those rules are, materialism and science are closely related (and occasionally confused with each other). The precise relationship between science and materialism is important for the present work, and we will examine it more carefully later.

Various strains of materialism have arisen in many times

and cultures. To get a better sense of materialist thinking, let's briefly examine five examples: materialism in Classical Antiquity, Enlightenment materialism, nineteenth century European materialism, the Carvaka school of materialism in India, and the American naturalist school in the twentieth century.

Among the Greek pre-Socratic philosophical speculations on nature, a version of materialism was prominent. The well-known "atoms" of Democritus were everlasting and indestructible, the only stuff from which all things are made. The forms and structures of the world, including ourselves and the gods, come and go as the atoms bundle together and fly apart; the atoms themselves remain unchanged. An important aspect of this thinking is that these atoms are governed by no kind of animating spirit or intelligence. They move, joining and sundering, only according to necessity, creating worlds and beings that pass away in time. Nothing else exists.

Democritus inherited these ideas from his teacher Leucippus, and the system was later incorporated into the broader philosophy of Epicurus. A more extensive formulation of the ideas is presented in the famous long poem by the Roman Lucretius, *De Rerum Natura* (*The Nature of Things*). But this school of philosophy was later eclipsed by the schools of Plato and Aristotle, involving Ideas and Purposes. It's all the more remarkable, then, what a modern-sounding ring the materialism of antiquity has, especially the idea that a deeper understanding of reality is to be found in the myriad motions of invisible particles. Just for this reason, we should also bear in mind the prescientific and highly speculative basis for the ideas, and the often wildly incorrect explanations of specific phenomena.

Materialist philosophies almost disappear in medieval Europe, reappearing in the Renaissance mixed both with Christianity and with the burgeoning new mechanical science.

The works of Gassendi and of Hobbes are examples of this reappearance, which is part of a more general phenomenon that we'll examine in more detail later, namely the desacralization of nature during and after the Scientific Revolution. For now, we'll look briefly at a purely materialistic strain of thought that resulted from these beginnings after about a century elapsed, during the Enlightenment period. The erosion of clerical authority accompanied an increasingly secular intellectual culture. A new form of materialism was developed, based on two pillars: advancing scientific knowledge (to which it was only loosely related) and a clear commitment to atheism (which had a polemical dimension). This version of materialism is presented in d'Holbach's *The System of Nature*. The atheism that had been latent in the materialism of antiquity now becomes a more overt and important point. Meanwhile, the link with science, though largely rhetorical rather than substantive, is an important innovation that continues on (in some sense) to the present day.

This association of materialism with science is even more strongly expressed in the writings of nineteenth century European (mostly German) materialists, such as Vogt, Moleschott, and Buechner. The rise of this materialist movement was driven in part by a reaction to the dominance of romantic and idealist philosophy and in part by the dramatic progress made in the sciences around this time. Two major scientific advances were particularly influential: the development of the principle of the conservation of energy and the explanation of the origin of species by natural selection. Energy conservation is fully consistent with the contention of materialism that nothing in the material universe is ever lost or gained, that the original uncreated material substance is only transformed. Natural selection, on the other hand, enabled thinkers to dispense with the previously perceived need for immaterial influences in the formation of living things. Buechner's popular exposition,

Force and Matter, was very widely read for decades, serving both as a vehicle to disseminate scientific knowledge and as a materialist manifesto. Interestingly, the promulgators of this strain of materialism were mostly scientists by discipline, not philosophers.

Materialism has not, however, merely been a Western philosophical tradition. Over two millennia ago, a materialistic philosophy known as the Carvaka school existed in India. Holding a doctrine called Lokayata, this school was a rival to the idealist and intuition-based Hindu schools of philosophy, in which the spiritual destiny of humans is of prime importance. In contrast, the Lokayata proclaimed "that only this world exists and there is no beyond. There is no future life. Perception is the only source of knowledge; what is not perceived does not exist. [...] The soul is only the body qualified by intelligence. It has no existence apart from the body."[35] Though the Carvaka enjoyed significant popularity at various times and places, it has in general been a minor influence in Indian philosophy, which has heavily emphasized introspection, the spiritual, a synthetic approach to religious tradition, and an aim toward non-attachment to worldly values. On the other hand, materialist tracts, featuring now the influence of Western science, were published in India during the twentieth century.

Our last topic here is also from the twentieth century, namely the American naturalist movement. The naturalists, represented by such thinkers as Dewey, Santayana, and Woodbridge, have been an important influence in intellectual culture. Their fundamental assumption is that all real phenomena and entities are part of an empirically perceptible natural order, and hence amenable to analysis along basically scientific lines. They are clearly materialists in the sense that they reject the reality of immaterial and transcendent beings and concepts (such as Platonic Ideas, or God and the human soul as conceived by many religions), but they are distinguished from other schools of materialism in several ways. For example, the American naturalists are not reductionist in their thinking; it is not their program to try to explain all things in terms of irreducible particles' behavior, and they don't necessarily assume this is possible. Also, the naturalists accept religion and a spiritual dimension of life (along with art, ethics, and so on) as real and important subjects for inquiry. Ultimately, however, their inquiry must be grounded in and limited by their fundamental assumption, which is materialistic in character.

Materialism and Science

Materialism as a philosophical position thus entails certain basic assumptions: matter (in a broad sense that includes both matter and energy) constitutes all that there is; nothing immaterial (including any possible spiritual order of reality) exists; as a corollary, only that which is accessible to the senses (and their extension through instrumentation) is real. This last point is important, because it connects materialism to science, since science makes the same restriction in its allowable subjects for discourse. We need to explore this relationship between science and materialism more deeply, but first we must clear up several points of nomenclature.

There is no consistency among different writers in their usage of the terms materialism, naturalism, and physicalism. Though sometimes distinguished by fine definitional differences, these terms are often used interchangeably in the literature. Since it appears to me that the use of any of these terms would satisfy the three basic assumptions just listed, my custom will be to use materialism as a broad concept that includes the basic meanings of the other two terms.

Some writers, however, restrict materialism to include only reductionist explanations of phenomena. There is no legitimate reason that I can see for this restriction. Whatever the merits or failures of reductionism as a philosophical doctrine, it is independent of our choice concerning a materialistic metaphysics; holism and emergence are equally consistent with the assumptions of materialism as defined above (not to mention being important in certain areas of modern science). Likewise, materialism is often assumed to imply determinism, but this usage is merely a relic of nineteenth century scientific thought. A strictly deterministic interpretation of phenomena is belied by both non-linear dynamics and quantum theory. Obviously, I will make none of these restrictions in the meaning of materialism used here.

There is also sometimes confusion between positivism and materialism, because positivism demands the elimination of unobservable entities, which would include the immaterial entities disallowed also by materialism. The two positions are not identical, however. Perfectly materialistic concepts might well be rejected by positivism if they cannot be tied directly to some empirical consequence (e.g. Mach's rejection of atoms). On the other hand, later versions of positivism (in order to avoid being criticized for making unjustified ontological claims) were carefully restricted to the logical analysis of language. Hence, later positivists would avoid making any explicit claims about a materialistic (or any other) basis for reality, the question being undecidable (and uninteresting) based on their philosophy. Positivism and materialism are indeed sympathetic to one another, but neither one actually implies the other.

This brings us face to face with our crucial question: how are science and materialism related to each other? That they are related is quite clear, similar in a sense to the assertion that positivism and materialism are related. "All scientists are formal materialists in so far as their philosophies can be deduced from their behavior."[36] In a broad sense, the mission of science is to study the properties of matter; if anything else has any claim to be real, then that lies outside the domain of science. But for this very reason, we can't legitimately claim that science implies some warrant for a claim of materialism. "If we take materialism in its nontrivial meaning, as the philosophical doctrine that denies the existence of nonmaterial realities, neither empirical science nor epistemology has anything to say about this. The reason is precisely because empirical science concentrates on the study of the material world and it makes no sense to derive from it assertions about spiritual realities. To interpret this as the 'ontology of science' is also meaningless."[37] The logic here seems very persuasive: surely we can't draw any conclusions from science about questions that are not within the purview of what science studies. This logic is correct, but oversimplified. Although the validity of materialism cannot be asserted based on any scientific findings, it is nevertheless true that science proceeds (methodologically) by implicitly presupposing that a materialistic worldview is correct. The illogic of making any materialist ontological claims based on this is beside the point; the relationship between materialism and science is deeply embedded in a thinking process, not a logical claim.

This relationship can be exploited by proponents of materialism who fully understand the logical issues. "Conceding that materialist philosophy as a whole is no more scientifically provable than its competitors, [Buechner] argues for its relatively greater plausibility on grounds of the greater conformity of its approach to reality with that of science."[38] Once this reasoning is accepted, the enormous success of science as an explanatory methodology and as a cultural influence may well incline many towards materialism, even those who nominally belong to some religious faith. Equally important, those of us who value science highly find ourselves, while doing science, in an ontologically comfortable materialist world even if we dispute the validity of materialism on other grounds. To repeat, we are not necessarily saved from this problem by the strictly logical compatibility of science as such with the philosophical rejection of materialism. I will argue that the solution to the problem lies elsewhere.

Materialism and Spirit

The type of materialism that we have been discussing so far does not attribute any properties to matter that are not discoverable by science. For this reason, I will call (in conformance with standard usage) this type of materialism "scientific materialism" and generally use the terms (with and without modifier) synonymously. However, it is possible to adopt a broader conception of the properties of matter that includes spiritual aspects, and this has been done by many thinkers from the ancient Greeks down to the present day. "In this conception spirit is chief among the internal properties of matter, and may accordingly be defined as the dominant inner quality of a material thing."[39] Such thinking avoids the problems of dualism while acknowledging the reality of a spiritual order by having the spiritual reality inherent in matter itself, as opposed to a lifeless and inert matter animated by a separate and immaterial spirit. Scientific materialism, of course, also avoids dualism by eliminating spirit altogether. The issues surrounding this view of matter as fundamentally imbued with mind and spirit, along with its relationship to science and to immaterial being, will be more carefully analyzed in due course. For now, I merely wish to maintain that in order to avoid terminological confusion such thinking ought not to be called materialism. In the present work, at any rate, "materialism" will always mean scientific materialism as we've defined it here. This scientific materialism is what I am calling the mundane view of nature.

This brings us to one last question: is an outlook of scientific materialism compatible with religion? If we restrict the scope of religion to ethical guidance or to personal meaning that we project onto the world, then the answer is clearly yes. This was essentially the position taken by the American naturalists. However, if we ascribe serious ontological reality to a transcendent ground of Being, for example, then our beliefs can no longer remain compatible with scientific materialism. These beliefs can certainly be compatible with science as such, but given the proposed linkage between materialism and science there is still a significant problem to resolve. A materialist worldview does not conflict with our *feeling* that nature is sacred, but rather with sacredness *being* a genuine attribute of nature itself.

4. The Sacred World

There is no single precise meaning for "sacred" when applied to nature. Many different cultures across many different times have interpreted the sacredness of nature in a variety of ways. In the midst of all this variety, though, we can identify a number of important commonalities. A perception of aliveness in nature; a sense of meaning; the infusion of a divine presence; an understanding of the interconnectedness among all things; and more particularly some special connection between humanity and the rest of nature; these are all typical elements of a sacred view of nature. Rather than trying to construct some dubious final definition, let's instead look at a set of illustrative examples taken from a wide variety of contexts, drawing heavily on a rich existing literature.

Before exploring examples, however, there is one theoretical point that we should address, namely the relationship between God and nature. One position is to maintain that God is identical with nature; this position is known as pantheism. Pantheism was the basis of many traditional cultures and is still popular in some circles. Pantheism automatically confers a sacred quality to nature, but it is also controversial in other circles and often seems to provoke bitter opposition. At the other extreme, God may be considered totally transcendent and outside nature. This view does not necessarily prevent nature from partaking in the sacred, however, because God may still bestow this quality upon nature from outside it. Between these more extreme views, there is the contention that God is both immanent in the world but also transcends the world, that God can be identified with but not limited to nature. This idea is sometimes referred to as panentheism (a position, incidentally, similar to the author's). Lastly, it is also possible for nature to have a sacred quality in the absence of any God at all, as some forms of Buddhism and as Taoism both demonstrate.

Let's now turn to illustrations of the commonalities found in sacred views of nature. The sense of nature as being somehow alive, as opposed to being inert, dead, and objectified, is one of the most commonly recurring themes in discussions of the sacredness of nature. This theme appears in a variety of variations and cultural contexts, combined with other elements that sometimes differ greatly from each other. The aliveness of nature is related to the interconnectedness of all things in the following example, taken from Native American religious philosophy: "What I am trying to communicate here is that the awareness of the sacred is an *experience*, not a hypothesis or some sort of demonstrated conclusion. But what is that experience of the sacredness of everything like? It is a way of seeing in and through and beyond each individual thing to perceive the relatedness and interconnection of each thing to everything else. Finding the sacred is not to have a theory about the origins of things or to apprehend their usefulness for us but to witness the miraculous *reality* and the unfolding *aliveness* of everything, a being connected together of all real entities. [...] This awareness of the mysterious and interconnected ground of being that is manifested in the eruption of each remarkable thing is the experience of the sacred."[40] In Greek nature conceptions, the aliveness of nature was often associated with the mind or soul of nature as a source of order in the cosmos. This idea is found in many forms in different philosophical systems over hundreds of years. As Collingwood puts it, "….the world of nature is saturated or permeated by mind…..They conceived mind, in all its manifestations, whether in human affairs or elsewhere, as a ruler, a dominating or regulating element, imposing order first upon itself and then upon everything belonging to it […] Since the world of nature is a world not only of ceaseless motion and therefore alive, but also a world of regular or orderly motion, they accordingly said that the world of nature is not only alive but intelligent; not only a vast animal with a 'soul' or life of its own, but a rational animal with a 'mind' of its own."[41] Of the several competing conceptualizations of nature found in the Christian traditions, some have also developed a sense of the aliveness of nature, in this case as warranted by God. Along with the famous love of nature shown by St. Francis of Assisi, one well known example of this approach is the nature philosophy expounded by Hildegard of Bingen. "Hildegard developed a theology that closely integrated God, human beings, and nature […] Central to Hildegard's theology is her notion of viriditas or 'greenness' […] It is vigor and fertility and health […] And viriditas is not simply a property of what we would normally think of as living things; it is also a property of such inanimate things as rocks […] The natural world is not, for Hildegard, simply inert matter, but is filled with life and power […] For Hildegard, the world still fit together like a vast organism, possessed of life, filled with greenness….."[42]

But aliveness is not the only attribute of nature construed as sacred. In fact, sacredness itself, in the sense of being sanctified or holy, is a primary quality not dependent on anything else. Consider these comments, written by a contemporary essayist just a few years ago: "But one did not have to be a naturalist to be awed by the sheer beauty of the animals and plants and rocks themselves, as they disclosed themselves to us in the here and now […] we felt God in the earth, not as a person but rather as a power. We felt Holy Wisdom through the sheer presencing—the suchness—of the bluff in its numinous energies. The suchness was itself a story, told not in words but in sheer splendor."[43] Another beautiful example of this sense of existence itself as being holy and sacred is found these words of the Irish poet and mystic who wrote in the early 20th century under the pen name AE: "I think of earth as the floor of a cathedral where altar and Presence are everywhere. This reverence came to me as a boy listening to the voice of birds one coloured evening in summer [...] and I felt a certitude that the same spirit was in all [...] So the lover of earth [...] will be tranced in some spiritual communion, or will find his being overflowing into the being of the elements, or become aware that they are breathing their life into his own [...] earth may suddenly blaze about him with supernatural light in some lonely spot amid the hills, and he will find he stands as the prophet in a place that is holy ground...."[44]

Another important dimension of sacredness in nature concerns the meaningfulness of nature, and the presence of symbolic content within the overt signs and occurrences found there. Appearances, in this view, serve as a gateway to the apprehension of deeper levels of meaning. "Nature includes not only what impresses the eye as color or form, but also inner dimensions [...] the world is no mere surface reality but a living cosmos that we can gradually learn to see...."[45] These deeper levels of meaning may be inherent solely within nature itself, or they may reflect another transcendent realm of which visible nature is a manifestation. Such deeper levels of meaning can be more real than the visible appearances themselves, especially in archaic and traditional cultures. For example: "Objects or acts acquire a value, and in so doing become real, because they participate, after one fashion or another, in a reality that transcends them [...] The object appears as the receptacle of an exterior force that differentiates it from its milieu and gives it meaning and value. This force may reside in the substance of the object or in its form; a rock reveals itself to be sacred because its very existence is a hierophany."[46]   Considerations of this sort, though, are not limited to the "primitive" cultures that Eliade was focusing attention on. The Romantic and Transcendentalist movements in literature embraced a similar kind of nature philosophy. In addition, many sophisticated civilizations throughout history have developed an essentially comparable world-view, sometimes in considerable detail. "Traditional metaphysics sees the universe not as a multitude of facts or opaque objects each possessing a completely independent reality of its own, but as myriads of symbols reflecting higher realities […] the light of the intellect, sacred in its own essence and also sanctified by revelation, penetrates into what appears as fact to reach its inner significance and meaning so that opacity is transformed into transparency. Phenomena thus become transparent to realities that transcend them […] Phenomena become gateways to noumenal realities."[47]

Let's look in more detail at two cases in which the aspect of meaningfulness takes on particular kinds of significance. One case involves the role of humanity and the other involves the role of the divine; we'll consider the role of humanity first. The relationship between humanity and the rest of nature becomes an important issue within the context of nature construed as sacred. In a mundane view, humans are simply a small subset of the natural realm and require no special category, but in a sacred world humans may acquire a special significance in the overall schema of nature. For example, "Hildegard believed the cosmos in its totality to be perfectly proportioned and balanced, and she believed that human beings reflected that perfection, precisely as the part of the cosmos that was capable of mirroring the totality."[48] Of course, the details of the role that humans play in the cosmos might vary greatly among different kinds of sacred conceptualizations, and these differences certainly need to be critically examined. But the fact that humans do have some special role to play is found frequently, and a corollary of this fact is the connection between the moral order in humanity and the natural order, a connection not found in a mundane world. Whether a particular version of this connection might be valid in a sacred world depends on the details of that version, but underlying all of them is the idea that there must be some genuine *relationship* to nature, and this idea is likely to be valid in virtually any sacred construal of nature. We see this in the following passage, also by a contemporary writer: "It is hard to feel a sense of gratitude for an inanimate, mechanical world proceeding inexorably in accordance with eternal laws of nature and blind chance. And this is a great spiritual loss, for it is through gratitude that we acknowledge the living powers on which our own lives depend; through gratitude we enter into a conscious relationship to them; through gratitude we can find ourselves in a state of grace."[49] Another point worth making is that humans can, through their relationship to and participation in the sacred dimensions of nature, enter into a transcendent state that surpasses the bounds of ordinary reality. Eliade has illustrated this point extensively for the context of archaic cultures. "Through the paradox of rite, every consecrated space coincides with the center of the world, just as the time of any ritual coincides with the mythical time of the 'beginning.' Through the repetition of the cosmogonic act, concrete time [...] is projected into mythical time, *in illo tempore* when the foundation of the world occurred."[50]

The role of the divine in bringing meaningfulness to nature also varies among different cultures. For the revealed monotheistic religions, God mandates whatever meaning and order nature might have. For example: "The order of nature is seen in the Islamic perspective to derive according to Divine Wisdom from the prototype of all existence in the Divine Order [...] The order of nature, therefore, reflects and issues from the order that exists in the Divine Realm [...] the cosmos is, symbolically speaking, like 'His garment,' which at once veils and reveals His Reality. The order of nature is not only created by God through His Will, but derives from the Divine Substance [...] the order of nature *is* none other than the Divine Order manifested upon the particular level of cosmic existence that we identify as nature."[51] In Buddhism, on the other hand, the *dharma* is an underlying source of order in both human and natural affairs. Likewise, the *Tao* plays a similar role in Chinese Taoist thought. These sources of order have no originator but simply are. In contrast, the monotheistic faiths attribute this role to God. These differences are real, but they should not mask the equally real commonalities. A crucially important commonality is that in each case something does play this role and functions as the source of order in virtually any sacred conceptualization of nature, and furthermore that the order of nature and the order within human affairs share this source.

To summarize, we have seen that a sacred view of nature is associated with a sense of the aliveness of nature, the inter-connectedness of all things, meaningfulness in nature, divine presence and/or other sources of order, and a special role for humanity in the cosmos or at least attention to the relationship between humanity and the rest of nature. Let's end with a few more examples that illustrate these various points. Many of our examples have been taken from descriptions of traditional cultures, but here is an excerpt from a modern essay on the metaphysics of nature: "Consequently, the cosmic organism, while it is one and indivisible, is at the same time a range of developing phases which can be represented and can display themselves as an evolutionary scale. The totality is constituted by the scale of its internal forms [...] each gives rise to the next higher level by virtue of the potentiality infused in it by the immanent principle of the totality [...] This is an idea of Nature, not merely as an all-embracing living animal, but as a dynamic organismic system [...] What we have so far maintained is that the universe is one single, indivisible whole, that it is self-specifying, self-differentiating and self-proliferating as a continuous scale of inter-dependent forms dialectically related."[52] This passage illustrates the themes of aliveness and interconnectedness without invoking any specific religious concepts associated with particular premodern cultures. The themes of divine presence and meaningfulness can also be found joined together in particular examples. The combination of these two themes can be seen, for example, in this excerpt discussing ideas taken from eastern Orthodox Christianity: "For St. Maximus [...] categories of order (*taxis*) in nature are established by God and denote God's active care for his Creation. The principles of Creation in its differentiated forms exist already in God. They are the *logoi*, which are the Divine Wills or Intentions. The *logoi* of various beings are held together in the Logos, which expresses the unifying factor in Creation whose purpose is a living relationship with God."[53] The details found in this particular religious context might vary from the details found in some different context, but the overall themes are found in many times and cultures. Let's end with a brief passage in which almost all of the themes we have examined throughout our discussion are implicitly found: "All things in the world and their totality are only a fragment of a still greater whole, σύμβολον of an inner world, and the rite takes on life only through the participation of the spirit."[54] A sacred view of nature, although not easily defined, can still be recognized and understood.

5. The Desacralization of Nature

Although various forms of atheism and materialism have existed in many cultures and times, I believe that the transition process that occurred in European culture from about 1400 to about 1900 is unique. Because the resulting cultural norms have spread throughout the entire world, greatly affecting (whether embraced or fought against) every culture that is part of the global communications network, the uniqueness becomes amplified. Despite the many counterattacks by fundamentalists, traditionalists, neo-pagans, deep ecologists, etc., I maintain that this mundane and secular view of nature is the dominant paradigm presently operating in the world. In this chapter, we'll explore in more detail how this situation came to be. Because the dominant religion in Europe during this time period happened to be Christianity, much of the historical narrative will be concerned with the relationships between Christianity, science, and the other intellectual currents operating during these times.

The Rise of the Mechanical Philosophy

The view of nature that existed prior to the process of desacralization had its own historical development. Interest in nature was always of lesser importance than salvation from the

earliest writings of the Church Fathers. But Greek learning
was known and valued by many of the early writers, and a
Christianized version of Platonic thought became standard
for many centuries. As long as it was rigorously subordinated
to Revelation and Scripture, the study of nature could be
encouraged. Such traditions were preserved by the writings
of those like Boethius, Cassiodorus, Isidore, and Bede at the
beginning of the Middle Ages, and developed slowly at various
intellectual centers such as the cathedral school at Chartres.
The development accelerated greatly during the 12th century,
as contact with the Islamic world brought a wealth of new
learning and greater familiarity with the Greek heritage. The
science of Aristotle, in particular, made a huge impression
on medieval Europe, and controversies ensued over the
compatibility of some Aristotelian doctrines (e.g. the eternity
of the world) and Christian dogma. The well-known ultimate
result of these controversies was the intricate and impressive
systematization of St. Thomas Aquinas, reconciling Aristotle
and Christianity into a synthesis that would dominant thought
for three centuries. This worldview is sacred in a variety of
ways. Perhaps most importantly, "the whole universe had been
created as the stage on which mankind would play out the great
drama of salvation."[55] Every item in the world is ordained by
the will of God, and has a purpose in God's plan. Although
the view of nature is more complex and sophisticated than
earlier views, it is still absolutely subordinated to Revelation
and Scripture. And although the interpretation of Scripture
is more nuanced and subtle now, there is still no question that
it is God's Word and unequivocally true. "This sense that
human beings are linked in harmony with the natural scheme
of things, and that they have their own distinct place within its
overall orderedness, made the medieval world picture a true
cosmology in the traditional sense of the term."[56] It is this
inherently sacred view of nature that would be challenged by

the novel intellectual forces stirring in the Renaissance.

One such challenger was the nexus of ideas that would eventually come to be known as the "mechanical philosophy." The seeds of the mechanical philosophy are found in both antiquity and within the Scholastic tradition itself. Despite the immense authority of Aristotle, scientific thought was advanced within the Scholastic tradition and in particular the study of motion was conducted in the 14th century by the so-called Paris Terminists. The main result of this effort, led by John Buridan, was the concept of *impetus*, which is a kind of power to move imparted to a body. This improvement in Aristotle's theory of motion served as a bridge to the development of modern theories in the 17th century. Another famous innovation of the Paris Terminists is the introduction of graphical techniques to study change by Nicole Oresme. These primitive beginnings were to bear much fruit later in the study of motion, central to the rise of the mechanical philosophy. But another latent idea from antiquity survived throughout the Middle Ages to achieve prominence later, namely atomism. The atomism of Democritus and Epicurus was preserved by the early compilers, along with other parts of the Greek heritage. It attracted little attention for many centuries, but eventually became a centrally important idea in the new mechanicism, and the atheism often associated with the Epicurean atoms was again unleashed into the intellectual environment.

The mechanical philosophy was not the only challenger of the Scholastic synthesis. During the 16th century, Scholasticism was attacked by humanists, Hermetic thinkers, Protestant reformers, and heirs to a revival of Greek skepticism. The humanists recovered Platonic and Neoplatonic thought, offering an alternative to Scholasticism, which they often considered an "adulterated" form of Aristotelian thought. The humanists also discovered cabalistic and alchemical texts, which along with Neoplatonism and natural magic merged to become the

Hermetic movement, another rival to the scholastic thinkers and sometimes the source of bitter attacks. Protestant thinking assumed many different forms, but often it emphasized piety and Scripture at the expense of reason making it unsympathetic to the rationalism embedded in Aquinas' system, which in any case would be suspect being part of Catholic orthodoxy. The rise of skepticism was a threat to many belief-systems at this time, and the predominant Scholastics undoubtedly offered a tempting target. The efforts of the humanists to recover, translate, and publish the texts of Sextus Empiricus helped instigate the skeptical current that culminated in the Enlightenment. Although all of these movements were enemies of Scholasticism, they were by no means united amongst themselves. Skepticism might be turned against alchemy, science, or even Christianity itself as easily as against Aquinas. Deep differences divided the Hermetic thinkers from the adherents of the mechanical philosophy, and the humanists often had little or no interest in mathematics and the sciences. Meanwhile, although attacked on many fronts, the Scholastic philosophers and theologians maintained hegemony over much of the educational system and could also deploy the vast coercive forces of the Inquisition against their enemies.

The Hermetic movement, although revolutionary compared to the Scholastic establishment, was arguably the repository of a sacred worldview. By the end of the 17th century, both Scholasticism and Hermeticism were fading from the mainstream cultural milieu, and the view of nature was entirely dominated by the mechanical philosophy, which we might term mechanism. A closer examination of how and why this happened, and to what extent this mechanism is inconsistent with a sacred worldview, will be our next major task. The *impetus* theories had been seamlessly incorporated into the Scholastic corpus with no problem, but in order to make further real progress in understanding falling bodies or projectiles, a more radical departure from the ideas of Aristotle was necessary. Most famously, the proposition that a falling body's speed depends on its weight needed to be shattered, a job first done by Philoponus in the 6th century using experiment and again, prior to Galileo, by the mathematician Benedetti using reasoning. More important than the specific fact was the general method, involving the use of mathematics, quantitation of the relevant aspects of the phenomenon, the use of simplifying idealizations and approximations (e.g. ignoring the influence of the medium), and increasing precision of measurements. In other words, a science of motion (kinematics) was in the making, and Galileo was both the premier practitioner and most effective promoter of the new discipline. We don't need to examine the details of the history of science too closely here in order to see the profound shift in methodology and attitude that was occurring. In Aristotle's physics, the crucial questions concern the causes and the essential natures of phenomena, all described verbally. Inspired by the mathematical studies of Archimedes and the role of mathematics in the philosophy of Plato, the reformers of Galileo's generation radically altered the very questions that were put to nature. "Is the whole of this geometrico-mechanical ideal world of forms […] a schematization of reality obtained by radical abstraction, in which a great many essential elements have been lost, so that it will never be able to furnish a reliable and complete knowledge of nature? Or is the world of sense-perception only a defective and imperfect representation in obdurate matter of the ideal realm of rational mechanics […] and should this realm be called reality in the proper sense? [...] Galileo misses no opportunity of expressing his metaphysical conviction that the structure of reality is essentially mathematical in character…"[57]

Meanwhile, another attack against Scholasticism came from a different direction, namely the explanation of the constitution of matter. A doctrine of substances whose essential nature changed during chemical reactions became increasingly difficult to reconcile with the experience of those who worked first-hand with such reactions. Some form of atomism seemed more consistent, and such corpuscular theories were revived in several quarters during the 17th century. An early and famous example is that of Pierre Gassendi, a Catholic priest who revived Epicurean atomism in a "Christianized" form to serve as an alternative to Scholastic philosophy. The specific changes required from the atomism of antiquity are easy to see: the eternal uncreated atoms need to be created by God, the deterministic necessity of the motions must be guided by God's providence, and an incorporeal soul must be added to the picture. Having made these changes to conform with faith, Gassendi believed that his otherwise mechanical world picture was the best way to understand phenomena; in particular, he believed that it was better than the Scholastic picture. His objections to Scholasticism mostly stemmed from its dogmatic quality, because Gassendi had been profoundly influenced by the skeptical revival. His early position was an extreme skepticism that precluded the possibility of having any knowledge at all, and he retained this attitude in his denial of the validity of any dogmatic systems, Cartesian as well as Scholastic. However, Gassendi eventually came to a position of mitigated and constructive skepticism in which some limited knowledge is possible in order to explain appearances even if the "true nature of things" can never be known with certainty. This "allowed him to formulate quite fully a scientific outlook devoid of any metaphysical basis, a constructive skepticism that could account for the scientific knowledge that we do, or can, possess, without overstepping the limits on human understanding..."[58] For Gassendi, then, atomism served as a useful hypothesis by which we can try to understand the phenomena we observe. Very few good explanations of this sort were actually available, however, given the limited information known at the time.

The greatest amount of precise measured empirical information at the time was found in astronomy, the last major contributor to mechanicism that we'll examine. By the end of the 16th century, the debates concerning Copernicus' heliocentric theory had become important and the extraordinarily precise measurements of Tycho Brahe were put to optimal use by Johannes Kepler. The brilliant success of Kepler in using mathematics, inspired by his Neoplatonic worldview, to finally find the correct orbital shapes for the planets (consistent with Tycho's data) is well known. Equally well known is the controversial and dangerous polemical success enjoyed by Galileo in his dispute with the Scholastics over heliocentrism, ending with his trial and recantation. Beyond the further erosion of the Aristotelian approach to physics and cosmology that a moving earth implied, the progress in astronomy also contributed to mechanicism in several other important ways. One way was the increasingly unavoidable association of some physically real interpretation of an essentially mathematical scheme. The reality of mathematical descriptions of nature, as opposed to their status as convenient calculating methods with no genuine meaning, was an important aspect of the mechanization of nature that eventually evolved. But another way that progress in astronomy contributed to mechanicism was through the ongoing search for some explanation of why the planets moved in the paths that were finally worked out, because the explanation that succeeded in the end turned out to be a great triumph for the new science of mechanics, namely Newton's Laws of Motion applied to the gravitational force problem. Even during the search for an explanation, the false starts and attempts indicate a slow trend toward mechanical explanations rather than the more organic and teleological approach used previously. For example, Kepler writes about replacing the word *anima* (soul) by the word *vis* (force) in his discussion relating planetary speeds and distance from the sun. "This appears to be no more than a substitution of one word for another, but the two words represent altogether different views. To read *vis* for an earlier *anima* is to abandon an animistic in favor of a mechanistic conception. As he expressed it elsewhere, Kepler wished to regard nature no longer *instar divini animalis* (as a divinely animated being), but *instar horologii* (as a clockwork)."[59]

Mechanicism and Religion

So we see that in the opening decades of the 17th century, at least three major intellectual currents are active that will form part of the mechanical philosophy: progress in developing the mathematical science of kinematics; the revival of atomism as a theory of matter; and the solution of the planetary orbital problem in astronomy. All of these endeavors deeply challenged the hegemony of Scholasticism and thereby challenged the sacred worldview that Scholasticism had built. This worldview had been accepted by the Catholic church as part of its orthodoxy, and so the challenge also extended to religious authority, for good or for ill. But did these ideas challenge religion *per se* or preclude any kind of sacred worldview? Our next task is to look more closely at the relationship between the beginnings of mechanicism and religion.

Perhaps the most famous response is that of Galileo, who proposed the idea of the two books given to us by God to find wisdom. One book is the Book of Scripture, or God's word. The other book is the Book of Nature, or God's works. The idea is that studying nature is an inherently religious activity since nature is the work of God, and the main confusion to avoid is looking in the wrong book; we ought not to look for signs of Revelation in nature, nor ought we to look for an understanding of how nature works by reading Scripture. In England, Francis Bacon devised an almost identical formulation at about the same time. Although this idea seems reasonable enough today, it had some controversial implications in the context of its time. Scripture was considered infallible, and if some passage should contradict the results of studying nature then the infallibility of the Word of God is called into question. For intellects like Galileo's or Bacon's, such contradictions are not a problem, because the Scripture can be interpreted allegorically or one can argue that its language is the imprecise language of the common uneducated person, and only the genuinely important spiritual truths need to be taken with complete seriousness anyway. But these sorts of arguments carried no weight with conservative theologians, especially at a time when Catholicism and Protestantism were locked in a deadly struggle over who is the legitimate interpreter of Christian doctrine. This issue of Scriptural interpretation became very important and we'll return to it later. Meanwhile, we need not examine the details of the political and bureaucratic infighting internal to the church that eventually resulted in Copernicus on the Index and Galileo on trial. The well-known outcome put Scripture on the losing side.

Kepler's views are less familiar but are extremely interesting. Although the previously quoted passage suggests that Kepler was aware of his contributions toward the mechanization of the world picture, his cosmological ideas were very complex and multifaceted. In contrast with Galileo, Kepler was willing to find actual spiritual messages in the study of nature. Kepler's conception of God was highly Pythagorean, a divine geometer whose mathematical creation shows up everywhere in the study of nature. More willing to speculate and apparently more actively religious than Galileo (who seems to be a good but somewhat disinterested Catholic), Kepler chose exile instead of religious compromise. In his writings, "he speaks of a divine frenzy, and of an ineffable rapture at contemplating the heavenly harmonies."[60] In addition to his mathematical and Pythagorean conception of God, Kepler also includes an explicitly Christian dimension and relates this also to his studies of nature. "Kepler's laws were never impersonal laws, however, and the universe in which they worked was not for him the chance product of their blind operation. It was an ordered cosmos consciously contrived [...] It was not by accident that Kepler's universe remained finite, finite and spherical, because to him the sphere represented an embodiment of the Trinity [...] He was prepared to inject the Trinity into purely scientific questions."[61] We see that Kepler is still living in a sacred world, despite his incipient enthusiasm for mechanical explanations. He sees nature as governed by mathematics, but this mathematical reality is mandated by a God who is still a living presence and who is simultaneously both Pythagorean and Mosaic. That a person living in this world was willing to give up his Platonic belief in perfect circular motion over a tiny numerical discrepancy with measured astronomical data is a testament to the genius of this enigmatic thinker.

A different approach to these issues is found in the case of our third major example, Gassendi. To understand Gassendi's ideas on religion and its relationship to mechanism, we must first recall that his starting point is an extreme skepticism. To a modern mind, this might suggest that he was an enemy of religious thought, but the opposite is true; Gassendi was part of an important movement in the 16th and 17th centuries, namely fideism. To a fideist, skepticism cast doubt on the evidence of the senses and cast doubt on the results of reason, but Revelation based on faith is beyond doubt and so faith trumps both reason and evidence. Christian dogma is not susceptible to skeptical doubt, but any sort of philosophical dogma is destroyed by such doubt. This position allowed Gassendi to postulate his formulation of Christianity-compatible atomism and then use the atomism scientifically to explain appearances within his system of constructive skepticism. There are, however, two dangers hidden within this fideist position. One danger is that God, although built in at the start, becomes irrelevant in practice. The atoms move identically whether they are eternal or created, "nor does it make much difference for the real course of events whether the atoms behave according to the laws of mechanics under divine conscious guidance"[62] or not. The second danger is that faith itself may someday become the target of an uncontrolled skepticism; once it has been unleashed, skepticism becomes difficult to maintain within desired limits. Historically, as we'll see later, an unintended consequence of fideism was indeed the erosion of faith as time wore on.

Development of Mechanicism

Meanwhile, the development of the mechanical philosophy continued. The most abstract and mathematical system was devised by Descartes, for whom experiment and empirical data were relatively unimportant. He believed that an entire physics could be achieved by reason alone, based on a small number of certain axioms. One of these axioms was that matter is identical with extension in space, a purely geometrical quantity. All of the so-called secondary qualities (taste, color, hardness, etc.) are banished as mere subjective states. Descartes goes on to elaborate a complicated scheme in which space becomes subdivided and forms itself into vortices of various sizes, thus introducing a kind of corpuscular theory into his ideas. He then outlines the various motions, impacts, and behaviors of these corpuscles under various conditions, occasionally making contact with the empirical world, and always subject to his fundamental axioms, such as the invariance of the total amount of motion in the world originally put there by God. Although much of his vortex theory was soon abandoned as fanciful, one of his axioms turned out to be crucially important: the state of motion or rest of a body never changes spontaneously but only due to the influence of another body external to it (this of course is what we now refer to as inertia and Newton's First Law). Descartes' system also has historical importance beyond its incorrect details, because it was the first serious attempt to produce a mathematical treatment of matter based on mechanicism, setting a goal for future generations. "Cartesian physics is [...] mechanistic in character. This implies that it uses no explanatory principles other than the concepts employed in mechanics: geometric concepts such as shape, size, quantity, which are used by mechanics as a department of mathematics, and motion, which forms its specific subject. It recognizes as actually existing in nature only those things which can be described and explained by means of these concepts [...] there is not a single difference between a running clockwork and a growing tree."[63] One reason for Descartes' influence is that he explicitly combined mathematics, science, philosophy, and religion into a grand synthesis. He tried to overcome skepticism in philosophy by finding an indubitable axiom to start with (the famous *cogito*) and make deductions from it, and several of his fundamental axioms involved acts of God. "Cartesian philosophy is quite inexplicable without God [...] He is the guarantor of right reason and of the reliability of clear and distinct ideas. It is God's immutability that ensures the existence of laws and nature and necessitates the conservation of motion in the world. It is God's continuous presence that conserves those laws."[64] But note that having been used initially to justify various axioms, God then disappears entirely from any of the deductions and we end with a system that is quite cold and lifeless, very different from the sacred cosmos of the Scholastics.

In England, one of the foremost proponents of the mechanical philosophy was Robert Boyle. Much of Boyle's work was in chemistry and alchemy, and he uses his stock of experimental results to show that the corpuscular conception of mechanicism offers a more consistent explanation than either Aristotle or Paracelsus can offer with their pictures. In Boyle's version of a corpuscular theory, which he avoids calling atomism, small fundamental particles form stable concretions, and these concretions in turn join together or subdivide during chemical reactions, accounting for the change from one substance to another. Boyle realizes that his ideas are at best hypothetical, since to specify any particular microscopic form or even to identify a specific substance as resulting from one of the elementary forms was beyond the ability of science in his day; he is actually proposing a research program rather than a grand system. But Boyle was also deeply religious and thought carefully about the implications of the mechanical philosophy. Some of his thinking is similar to what we've seen in Galileo and Gassendi: The study of nature reveals an aspect of what we can know about God, and God both imparts and preserves the motions of the particles that make up the world. But Boyle also adds some new elements to the picture: "God has also revealed himself in nature, and the intellectual contemplation of the universe [...] leads to the recognition that the world must have been produced by an intelligent cause [... Boyle] opposed in particular the deistic view, which considered God's collaboration in the preservation of nature superfluous..."[65] Boyle goes on to include a discussion of Christian revelation, but the assertion of the ongoing workings of God in nature to produce the world we live in, and the consequent introduction of design into the discussion, are noteworthy. These themes continue down to the present day, and they became especially prominent in English thinking throughout the 17th and 18th centuries.

The culminating event for the mechanical philosophy, which insured its domination of intellectual activity for centuries, was the work of Isaac Newton. Newton followed Boyle in his general adherence to a corpuscular view of matter, but his greatest work was in the tradition of Galileo and Kepler, developing a mathematical science of mechanics and applying this to the planetary orbital motion problem. Hence, these two problems (dynamics and astronomy) merged into a single problem, which Newton solved brilliantly. The science of motion had now been placed on a rigorous basis, and it began to synthetically incorporate a variety of other sciences (acoustics, statics, hydraulics, pneumatics, thermodynamics) within it over the years. This was the golden age of classical physics, and it was all based on a mechanistic picture. This mechanicism was not, however, identical to the old mechanical philosophy of the early 17th century, because the forces between bodies were not merely contact forces but were instead forces acting over distances in some mysterious way (gravity being the first and most famous example). Although controversial at first, such forces soon became a routine part of the thinking process in classical mechanics and then in electricity & magnetism. After Newton's synthesis, the mechanical world picture had no serious rivals. Not surprisingly, the metaphor of nature as a machine or a clockwork soon came to be a dominant cultural image.

Implications of Mechanicism

Nature as a machine, then, had come to replace nature as an organic and sacred cosmos. But doesn't the machine also have some sacred dimension to it? After all, Gassendi, Galileo, Kepler, Descartes, and Boyle had all invoked God in the construction of their world machines. Newton also had religious convictions, and his identification of the existence of absolute space with God along with the necessity for God to maintain the gravitational attraction between bodies (which inert matter alone could not do) were both important parts of his thinking. He believed that these were good arguments against atheism, and his contention that God continuously acts to maintain the world is inconsistent with deism. Many thinkers in Newton's time (apparently with his approval) and afterward used these ideas to promote a religious vision of the world. "Rational analysis of nature, they held, showed that the world cannot be explained by natural causes alone but must be the work of God. Newtonian science demonstrated the need for a Creator, who graciously sustains a world totally incapable of sustaining itself [...] The mechanical worldview, in their hands, became an uncompromising witness to the glory of God."[66] On the other hand, this "natural religion" has nothing specifically Christian about it, nor does its God have any personal or caring relationship with humans. All of these thinkers *declared* the existence of God in their mechanical systems, but did any of these mechanical systems really *need* God for them to run?

Even while the mechanical philosophy was still struggling against its Scholastic and Hermetic competitors, some thinkers were already suggesting that it implied atheism. "This appears most clearly of all in the purely materialist metaphysics of the English philosopher Thomas Hobbes, who deduced not only the processes of inorganic nature but also all mental phenomena from matter and its motions, and consequently denied the existence of immaterial substances. And even if his works had not existed, the anxiety about the influence of mechanicism on religion, voiced by several seventeenth-century theologians and philosophers [...] would be evidence enough that the materialist metaphysics was gaining ground."[67] Reinforcement of this trend came from another quarter, namely the continuing development of the skeptical revival. The arguments of the fideistic apologists for both the Catholic and the Protestant sides in the religious conflicts had backfired. Each side had accused the other of having no grounds for its beliefs, using the tools found in the skeptics' arsenal. Each side was serenely confident in the rightness of its own beliefs, based on a faith that was beyond the reach of skeptical criticism, but it doesn't take much thought to see that if both sides cannot possibly be right then it may well be that neither side is right, and there are no grounds for believing Christianity at all. These attacks based on skeptical arguments "appeared to have a peculiar recoil mechanism that had the odd effect of engulfing the target and the gunner in a common catastrophe."[68] Ultimately, this sort of skeptical thinking eroded the faith that was becoming the only influence left welding the mechanical and Christian worldviews together. Even among those who professed to have no doubts, their need to profess so demonstrated itself the very existence of those subterranean doubts. "Everyone is familiar with the religious proclivities of scientists [...] in the late seventeenth century. The period has been called the golden age of natural theology, and with good cause. Never has atheism been so roundly refuted. One scientist after another took up his pen to demonstrate the existence of God, and [...] they called upon the latest findings of science to frame their arguments [...] I contend nevertheless that the meaning of it all is the secularization of nature, by which I mean to say that the scientific revolution effectively severed the cord that had bound the study of nature to Christianity during the previous millennium. The natural theology appears to me as a manifestation of the secularization rather than its denial, the effort of men uneasily aware that the ground was shifting under the traditional foundations of Christianity to construct new ones."[69]

During the course of the 18th century, the potential for decoupling the newly triumphant mechanical picture of the world from its religious overtones was increasingly realized. Within a generation or two after Newton, the discourse tended to deemphasize the role of God that Newton himself had found so important, and thinkers like Taylor, Folkes, and Pemberton slowly fostered the association between Newtonian mechanics and Deism, in part as a bulwark against the more radical claims of materialists and atheists. "The assimilation of Newtonian science into Western thought produced the first generation of

European thinkers for whom faith in the order of the universe proved more satisfying than faith in doctrines, creeds, and clerical authority."[70] But since Deism provides no ongoing role for the action of God in the world, it then becomes a short step to the elimination of God altogether. During the middle of the 18th century, working in the context of Enlightenment thinking, d'Alembert and Maupertuis argued that no religious inferences could be legitimately drawn from a scientific exploration of nature. La Mettrie saw no need to retain the extraneous elements that had been added to mechanistic pictures, and he argued that the order observed in the world had no need of a God to explain it; he even believed that human life and human reason were susceptible to a purely mechanical understanding. "Eighteenth-century philosophers generally agreed that nature should be studied directly without reference to a divine Being whose existence could not be ascertained and whose intervention into natural phenomena put a limit, so to speak, on the power of human understanding. The eighteenth century thus witnessed a general tendency to transfer to Nature all the powers traditionally attributed to God."[71] For example, the gravitational attraction that Newton had seen as God's ongoing and direct working in the world became simply another property of matter.

The culmination of these trends, and an exemplification of the way that the success of classical mechanics as a purely mathematical and scientific discipline lent credence to them, is found in the work of Laplace. Laplace is perhaps most famous for the anecdotal account of his reply to Napoleon's query about the absence of God in Laplaces's work ("I have no need of that hypothesis"). In physics and mathematics, however, Laplace is well known for a number of elegant and important studies and techniques. Two of his pieces of work had direct implications for the idea that God actively intervenes in the mechanical affairs of the world, because Laplace solved two puzzles that Newton himself had suggested as evidence of divine intervention. One problem was the apparent instability of the solar system, which Newton had speculated might need God to tweak the motions a bit at some future time in order to restore stability. Laplace, in a brilliant mathematical analysis, demonstrated that the apparent instability is actually a small segment of a long-period stable oscillation. Another unanswered question for Newton was why the planets all orbited in the same direction and in the same plane, unless God had initially set them into such a motion in the distant past. The celebrated "nebular hypothesis" of Laplace, which he worked out in great detail, provided a purely mechanical (i.e. physical) explanation for these facts. "Newton had used the small likelihood that random chance was responsible for the peculiar arrangement of the planets and their satellites as an argument for belief in Divine Providence. Now Laplace was proposing a reasonable mechanism for these features of the solar system, shortcircuiting the need for God as the direct cause of its peculiar configuration. It was especially for this invention that Laplace was repeatedly taken to task later for being a materialist and an atheist."[72] But Laplace did not promote any kind of theological view in his work, atheist or otherwise. The importance of Laplace for our present purposes is his works are purely mathematical and scientific; religious issues are simply not discussed at all. That he can, and does, do this is the important point here.

Laplace's work illustrates the astonishing success of the mechanical world picture perfected by Newton, and his work was matched by many similar successes in celestial mechanics by other investigators, as well as equally striking successes in applying the theory to other problems (the vibrating string as the basis of musical acoustics and the kinetic theory of gases are two excellent examples). In addition to the Newtonian program, the atomic constitution of matter was put on a more rigorous basis by Dalton and Lavoisier (the kinetic theory of gases also contributed here), confirming the other strand of the old mechanical philosophy. In other words, mechanicism was a strikingly successful paradigm, and to the extent that mechanicism lead to a picture of the world devoid of any sacred qualities, its continuing string of dramatic successes continued to desacralize nature.

Biblical Criticism, Geology, and Agnosticism

But a more direct attack on the traditional and sacred view of the world was coming from a different quarter. As the skeptical revival reached its high-water mark during the Enlightenment, the Bible was studied as an essentially secular document, as a text like any other text subject to historical scrutiny and critical hermeneutics. This idea had its roots in the work of the humanists and skeptics in the 16th and 17th centuries, when Erasmus discovered that certain important texts of the New Testament did not appear in the earliest available manuscripts. Critical Bible scholarship was performed by La Peyere and Fisher around the middle of the 17th century. "Exploring problems of ascertaining the correct text of Scripture and the status of Scripture as a source of truth, it raised many problems, some that became very important in creating doubts about religious positions."[73] The problems could be quite crucial, since Scripture was the only source of legitimacy for many Christian denominations. These early efforts were rather easily suppressed, though, despite their influence in limited scholarly circles. By the middle of the more secular 18th century, the effect of subjecting the Bible to scientific study became devastating. The word of God had once been declared an equal partner to the works of God found in nature, but now the Bible was just one more object in the environment to analyze objectively. Under these conditions Scripture could not maintain its authority, and the authority of Scripture was the foundation of the authority of Christianity itself. This process was mostly undertaken in German academic circles, where a 1779 publication by Eichhorn "marks the beginning of the modern critical movement"[74], and it spread from Germany to England and the United States during the 19th century. The infallible Revealed Word of God became replaced by questionable historical authenticity, textual comparisons, disputed translations, and interpretative quibbling. The only escape from this situation was the wholesale rejection of modernity.

The use of biblical authority, as a warrant for a sacred view of nature, suffered another major blow during the 19th century as a result of the arguments about the age of the earth. An early form of geology began during the 17th century with the work of Steno, who was the first person to propose that the surface of the earth tells us something about its history. Steno also first suggested that fossils are the remains of organic forms and he worked out the geological history of Tuscany from his observations. Interestingly, Steno found no discrepancies between his conclusions based on observation and the Scriptural accounts, since he had no independent way to estimate the ages of the rocks. As methods of dating rocks began to develop, however, geologists became convinced that the age of the earth was far greater than that indicated in Genesis. By the late 18th century, the geological strata were interpreted as the result of a long slow process, even if precise amounts of time could not be given. A consensus formed that the age of the earth must be extremely great, and some workers such as Hutton proposed that it had no beginning at all. "The earth, it was claimed, had always been and would always be under the dominion of the same purely natural laws."[75] Whereas the earliest efforts to reconstruct geological history were intended to supplement the Genesis narrative by supplying a few details left out by Moses, the relationship eventually came to be reversed; the geological reconstructions were taken to be a factual record and the burden was upon Christian commentators to show that the Genesis narrative could somehow be reconciled with it. This relationship was not necessarily anti-religious, since it fell within the traditions of natural theology, but it clearly undercut the authority of Scripture and it could be used effectively as a weapon against religion by those inclined to do so. Thus, a conflict between the age of the earth deduced from geological evidence and the age of the earth given in the Bible could be exploited both by defenders of orthodoxy and by the heirs of the Enlightenment for their own ideological purposes. These ideological purposes, however, whether Christian or anti-Christian, had little connection with the ongoing detailed work of geologists. "Toward the end of the eighteenth century many naturalists engaged in the study of the earth tried to dissociate themselves from the use of their work by rival cosmological interests [...] a firm rejection of large-scale theorizing and a new emphasis on the value of cumulative observation [...] expressed an attempt to establish the study of the earth as a practical pursuit that would be free of cultural pressures from *either* side: from the traditional concerns of biblical chronologists *and* from the secularizing concerns of eternalistic theorists."[76] This situation proved to be unstable, however, because during the course of the 19th century the results produced by the highly successful new scientific approach in geology continued to contradict the biblical narrative ever more reliably. In addition, the new sense of professional identity and professional autonomy of the geologists, promoted by the work of Lyell, served to marginalize and invalidate the "Mosaic geologists" and made the discipline more secular by default even in the absence of any ideological agenda. The stakes here were very high, because the Genesis narrative was central (for Christians) to a vision of humanity as having a home in a sacred cosmos. While this traditional vision might still be clung to with some effort in a world seen as mechanistic, the new paradigm of earth history could only be reconciled with the Genesis narrative by massive efforts of accommodation.

Another aspect of the new geological understanding was the interpretation of the fossil record as a chronological documentation of increasingly complex life forms, or what we would now call evolution. Although such a progression of life forms contradicted a literal reading of Genesis, it was still consistent with the tradition of natural theology. "Geology, or more particularly the analysis of extinct organisms preserved as fossils, gave a new temporal dimension to the sense of divine design in the world. The traditional static concept of design was dramatically enlarged by the understanding that divine providence had underlain equally all the successive phases of earth history, even before the existence of mankind..."[77] The facts were equally consistent with either a naturalistic interpretation or divine providence. For this reason, the mechanistic explanation of evolution proposed by Darwin (i.e. natural selection) was another major blow against the possibility of retaining some sort of sacred view of natural history. The subsequent linking of Darwin's theory with a naturalistic ideology by Huxley and other supporters of both Darwin and a secular worldview, the aggressive promotion of these views, and the polemical fights with conservative Christians and biblical literalists are all well known. Thoughtful scientists who were devout Christians, such as the botanist Asa Gray, were able to synthesize their faith with Darwinian theory, but the prevailing currents in the history of thought were against them. "...by 1900 the public stance of scientists in the English-speaking world (including some who followed Gray in botany) had overwhelmingly come to resemble the views of Huxley."[78]

By the 20th century, then, the desacralization of nature is essentially complete, at least in the Western world. In addition to the triumph of Huxley's agnosticism, we find the rise of logical positivism in philosophy and a somewhat uncomprehending faith in technology throughout the popular culture. The dominant influence in higher education had shifted from the Christian clergy to an alliance of humanists influenced by the German academic model and scientists influenced by naturalism. Although the revolutionary developments in relativity and quantum theory dethrone the old Newtonian classical formulation of mechanicism, the truly crucial aspects of mathematical and lawlike behavior in nature are only reinforced by these new ideas, and they do nothing to change the desacralized status of nature. Religions make very few cognitive claims concerning nature, and are sometimes even consigned to the limited role of offering ethical advice to guide human behavior. Even this limited role was usurped by the claims of both behaviorism and sociobiology, though such claims have been highly controversial. Attacks by fundamentalists on the hegemony of secular culture, though sometimes politically potent, are not taken seriously by intellectual elites and wouldn't restore a sense of the sacred in nature even if they succeeded. Attacks by postmodernists on the hegemony of a scientific worldview are only made within a broader still-secular cultural context, and have been equally futile in restoring a sense of the sacred in nature. To restore to nature its status as a meaningful and sacred cosmos will require a cultural shift as wide-ranging and radical as the one that occurred in the 17th century transition to the mechanical philosophy, and I don't believe that we can abandon what we have now achieved but instead will need to build upon it.

Part II

The Central Idea

6. Quantum Mechanical Background

Historical Synopsis

Science in the year 1900 seemed to be on the verge of achieving a complete understanding of the natural world. Optics and electromagnetism were not only complete but had been unified into a single theory by the work of Maxwell. Newton's mechanics had been vindicated spectacularly by ever more stringent tests over a period lasting more than two centuries, unifying disparate areas such as acoustics and the study of fluids under a single theoretical structure in the process. While there was still much to learn in sciences like biology and geology, a fundamental basis in chemistry had been proposed for these sciences, and chemistry itself had been put on a quantitative footing similar to that of physics; indeed, chemistry and physics themselves were coming closer together in fields like thermodynamics and electrochemistry. A unified understanding of all things, based on existing knowledge, appeared to be within reach.

As is now well known, the scientific knowledge existing in 1900 had to be radically revised during the course of a series of scientific revolutions (quantum mechanics, relativity theory, plate tectonics, molecular biology, and so on). We will focus our attention on what was arguably the most revolutionary of all these changes from an epistemological viewpoint, namely quantum mechanics. And the beginning of the quantum revolution occurred itself in the year 1900, with the introduction of the so-called "quantum of action" into physics by Max Planck.

In essence, the problem that Planck was trying to solve was a straightforward one. He was trying to explain the emission of radiation from a hot object (such as a red-hot poker, light from an incandescent bulb, or the infrared rays that night vision goggles detect). There was good measured data showing how the amount of radiation depended upon temperature and frequency, and Planck wished to explain these dependences. He applied standard thermodynamics and electromagnetic theory to analyze the problem, but the results of these well-established methods failed to agree with the data. In desperation, Planck altered one of the fundamental assumptions of the method, namely the assumption of continuity in the energy of the system. In other words, he stopped allowing the energy to have any possible value, but instead restricted the energy to a discrete set of particular values. More specifically, the energy needed to be directly proportional to one of the allowed frequencies of radiation in the system. Only energies meeting this condition could exist now in his analysis. In the mathematics of his method, he needed a constant of proportionality between the energy and the frequency. This constant, designated by the letter **h** and now called Planck's constant, turns out to be a fundamental constant of nature, deeply important in the structure of physical reality.

When Planck introduced the quantization of energy (i.e. dividing it into discrete non-divisible amounts) and Planck's constant into physics, however, their significance was not realized. Planck and his audience assumed that this was merely a methodological trick, an *ad hoc* assumption that brought agreement with experiment but that had no genuine physical meaning. The situation changed in 1905, when Einstein also proposed that the energy of light comes only in discrete amounts equal to **h**f (Planck's constant times the frequency of the light) in order to explain the experimental data of the photoelectric effect (ejection of electrons from a metal when light shines on it). Einstein's explanation was dramatically verified by the careful experiments of Millikan. We now needed to take seriously the idea that light energy is quantized, because the data unambiguously demanded it. Other kinds of experiments also supported this conclusion. Names for these packets of energy, such as "light quanta" and "photons," were coined. The photons behaved, in physical interactions, like small particles of light. But these results implied a terrible conceptual difficulty.

To understand this conceptual problem, consider the history of our understanding of light. In the seventeenth century, rival theories of light as a stream of particles ("corpuscles") and of light as a wave were both proposed (by Newton and Huygens, respectively). Various forms of evidence and reasoning were offered on both sides, but during the nineteenth century it became clear that light was a wave. Centering on the work of Young and Fresnel, huge amounts of empirical verification, in the form of sophisticated optical wave interference experiments, left no doubt that light was clearly a wave phenomenon. Maxwell's synthesis completed this work by discovering the fundamental nature of light as an electromagnetic wave. Hence, by 1900 it was a well known physical fact that light is a wave. None of this work or understanding was invalidated or rendered obsolete by the subsequent discovery that light is a particle.

The magnitude of the conceptual problem posed by these results is emphasized by a consideration of what we mean by "wave" and by "particle" in physics. A particle is a single isolated entity, virtually impenetrable and almost pointlike in extension. In contrast, a wave is a disturbance (i.e. not itself an object) in some medium, spread out through space ("spatially delocalized"). Although both waves and particles can carry energy and momentum, they are totally different physical conceptualizations. The issues involved are starkly illustrated by the fact that the energy of a photon (light particle) depends on the frequency of a light wave; frequency is a property of waves, it is meaningless to speak of the frequency of a particle. The energy of a light wave, meanwhile, is completely independent of its frequency in classical electromagnetic theory. And yet, all of these conclusions are inescapable results of experiment: light is a wave and light is a particle, and waves are not particles. Physicists were trying to understand this paradox in the early part of the twentieth century.

The situation then became even worse. Reasoning by analogy with the case for radiation, deBroglie proposed that particles with mass (such as electrons) also have a wavelike aspect, and he developed a relationship between the momentum of the particles, the wavelength of their associated waves, and Planck's constant **h**. This proposal was put to an experimental test (albeit by accident). When beams of electrons were directed toward the periodic array of atoms in a crystal, the results showed wave interference effects just as x-rays (a short-wavelength form of light) did. The wavelengths of the electrons can be measured in this experiment, and the measured wavelengths are just as deBroglie had predicted. Thus matter as well as light had now shown itself to be composed of both waves and particles, deepening the paradox.

Another strange result emerged from the work of Niels Bohr in 1913. Bohr was trying to explain the so-called spectral lines of the hydrogen atom. After energy is imparted to an atom (by an electrical discharge, for example), the excited atom relaxes back to a lower energy state by emitting light. This light consists of a discrete set of frequencies (different frequencies have different colors), and light of each different frequency can be deflected through a different angle by an instrument called a spectroscope. In this way, the different frequencies can be separated and measured; the resulting set of measurements is known as a spectrum. In the case of hydrogen, these frequencies have a simple mathematical relationship to each other, named the Balmer series after the person who discovered it. So, Bohr was trying to develop a model for the atomic structure of hydrogen that could predict the frequencies of the Balmer series. Astonishingly, he discovered that he could accomplish this by postulating that the angular momentum of the electron orbiting the proton in hydrogen was quantized in multiples of Planck's constant. From this followed the quantization of the orbital radii and orbital energies. Each different orbit corresponded to a different state of the atom. Bohr's method was another *ad hoc* procedure with no reason to be correct based on classical physics, but it agreed with the experimental spectra to a very high degree of precision.

In the Bohr model of the atom, the energy of the atom can only change discontinuously as it jumps from one quantum state to another; no other energies are allowed. The energy difference between the atomic states is then equal to the energy of the emitted photon. It is this quantization of physical properties such as energy in the atom that is radically non-classical, and once again the quantization is related to **h**. Although these various quantization processes postulated during the early work in quantum theory were not logically coherent, they clearly were pointing toward some correct and important aspect of physical reality.

After about another decade of further experimental and theoretical work, these early efforts culminated in a more rigorous formulation of quantum theory, which is essentially the same theory mathematically that we use today (for non-relativistic cases). This last step consisted primarily of two major contributions, one by Schrodinger and the other by Heisenberg. In Schrodinger's version, the wavelike aspects of matter are emphasized and the theory provides a method by which one can calculate the wave amplitudes of the system for some given potential energy function. Observable properties, such as the discrete set of allowed energy values, are also calculated in this method and can then be compared to experiment. The physical meaning of the wave amplitudes themselves, however, does not correspond to any observable classical variable; instead, the square of the wave amplitude corresponds to the *probability* of the associated particle's position. Hence, we no longer have a strictly deterministic physics. In Heisenberg's version of the theory, pictorial models of the microscopic world are abandoned altogether and the theory instead concentrates only on the calculation of observable properties (like the energies) using a mathematical method involving matrices. Once again, though, probability intrudes into the results, because it turns out that not every observable variable can be calculated with exact precision. In fact, although they employ different vocabularies and algorithmic methods, these two formulations have been shown to be mathematically equivalent to each other.

Determinism, Probability, and Uncertainty

The role of probability in quantum theory is crucially important, and it has equally important epistemological consequences. The goal of physics, for hundreds of years, had in some sense been the precise determination of the state of a system's motion, including all of its important dynamical variables (energy, position, momentum, angular momentum, and so on). We now have a theory that does not tell us where we will find a particle at a certain time, but rather the likelihood of finding it in various places. Related to this fact, the precision with which we can predict and/or measure certain variables is limited. The most famous (but not the only) example of this kind of restriction on our knowledge is the well-known Heisenberg Uncertainty Principle. The Uncertainty Principle states that the product of the uncertainties of some pair of variables, such as position and momentum for example, can never be less than $\mathbf{h}$, Planck's constant (more precisely, it's never less than $\mathbf{h}/4\pi$). The "uncertainties" referred to here are not merely measurement errors, despite the suggestiveness of the often-used gamma ray microscope example. These uncertainties are fundamental restrictions on what we can know about the system, limitations on our knowledge built into the very structure of the theory. The standard interpretation of the theory goes even farther: the Uncertainty Principle represents a restriction on *what there is to know* about the system. In other words, the theory implies that simultaneous exact values of the position and momentum of a particle do not exist.

The importance of these results derives in part from the way in which determinism is formulated in classical Newtonian dynamics. The relationships between position and momentum, as they change with time, are governed by a fundamental equation ("Newton's Second Law"), and this equation predicts all future values of these variables if we know their present values (plus all forces acting on the system). This ability to predict is essentially what we mean by classical determinism and causality. To be told that we cannot know, in principle, the simultaneous values of position and momentum for a particle strikes at the heart of this formulation. Determinism and causality have not disappeared from quantum theory, because the time development of the wave amplitudes is perfectly predictable; however, in order to convert the wave amplitudes into some measurable results, chance and probability must be invoked. These aspects of quantum theory are the basis of Bohr's frequently stated point that a space-time description of nature and a causal description of nature are no longer compatible. The use of one description automatically rules out using the other.

At this point, I wish to emphasize that these conceptual conundrums have not prevented the stunningly successful application of quantum theory to a wide range of physical problems for over three-quarters of a century. Quantum theory can explain the workings of a laser, conductivity in metals and semiconductors, chemical bonding, radioactive decay, the behavior of liquid helium, the fusion reactions that power the sun, and the operation of the solar cells that use the sun's rays to produce electricity for us. The numerical accuracy of the theory's predictions has been verified to small fractions of a part per million. All of our electronics, information, and telecommunications technology is based upon our understanding of quantum theory. There is no doubt that the theory is correct. The only controversy concerns how we should interpret its epistemological message.

This controversy intensified in 1935, when Einstein, Podolsky, and Rosen proposed a thought experiment to demonstrate that the theory is "incomplete" in the sense that it can't provide values for genuine physical variables that exist. Their idea is based upon a peculiarity built into the structure of the theory, and is related to problems with the measurement process. Since the theory only gives us a probability for the state of the system and hence the value of some variable, but a measurement gives us the actual value (i.e. which of the various probable values is the "true" value), then we can say that the measurement process "chooses" in some sense which state the system is in. This problem in itself is bizarre enough, but the EPR thought experiment sought to demonstrate that the measurement problem results in a paradox. The idea is this: Suppose some particle decays into two new particles, and these two particles are insured (by a conservation law acting during the decay) to have absolutely correlated values of some variable.

Before the measurement, we don't know the value of this variable for either particle. After we measure this value for *one* of the particles, we then automatically and simultaneously know what the value is for the *other* particle, even if they are now separated by a large distance. But, this kind of instantaneous action at a distance is forbidden by relativity theory, and according to EPR could not be characteristic of any "reasonable" view of physical reality. For a while, the arguments were purely epistemological; but three decades later, Bell showed that there are empirically testable differences between the orthodox quantum theory interpretation and any kind of "locally real" theory, and two more decades afterward the technology existed to do the required experiments. When the experiments were done, the results were consistent with the predictions of quantum theory.

Let's end this discussion by emphasizing once again the fundamental root of the issues, namely the existence of Planck's constant **h**. If **h** were equal to zero, then we would be able to know the values of all variables with infinite precision (in principle if not in practice). The determinism and causality of classical dynamics would hold true also in the microworld. Indeed, the reason that we usually don't perceive any quantum effects in our normal activities is that the value of **h** is very small (about 10^{-36} using our standard units of meters, kilograms, and seconds), and thus it can be taken to be approximately zero for such cases. But it's not zero. When we look on a fine enough scale, the non-zero value of **h** becomes crucially important, introducing a fundamental discontinuity into our interactions. Ultimately, it is the existence of this discontinuity (what Bohr refers to as the quantum postulate) that is responsible for all of the weird and paradoxical aspects of quantum theory.

7. Bohr and Complementarity

The Need for an Interpretation

After the work done by Heisenberg and Schrodinger, the physics community possessed a self-consistent mathematical formalism that seemed to predict the results of existing experiments. But what did the formalism mean? What did it tell us about reality in the microphysical world? Important parts of the theory, such as the wavefunction, had no clear physical meaning. Paradoxical results like wave-particle duality still remained. The implications of the Uncertainty Principle were an unsettled and unsettling issue. And unlike any of its classical physics predecessors, quantum theory delivered probabilities in place of exactitude. Despite the undoubted success of the mathematics, the theory could not really be considered a completed achievement. Some physicists believed that a new (or at least radically improved) theory was needed, but Niels Bohr was convinced that quantum theory was already revealing important truths about nature; what was required, he believed, was a proper framework within which to *interpret* the new theory.

By interpretation, we mean that each symbol in the mathematical formulation has a well-defined and unambiguous physical meaning, and that all the content of the theory corresponds in a clear way to physical reality (as revealed by the results of experiment). Given the success of the mathematics, a suitable interpretation would be tantamount to a complete and successful theory. The difficulties encountered in searching for such an interpretation, however, turned out to be horrendous. Bohr, who had been grappling with these issues for more than a decade, was almost driven to despair by his feverish efforts to overcome these difficulties. In the end, he found that he needed to reexamine and redefine the very concept of what a physical theory ought to be. He first presented these far-reaching conclusions at a 1927 conference in Como, and continued to refine and extend the ideas until his death.

We will examine the success of Bohr's interpretation, along with the attacks made upon it and the rivals it now competes with, in due course. We will also examine very carefully the important concept that lies at the heart of the interpretation, the concept of complementarity. But first let's develop the argument for Bohr's ideas in much the same way that Bohr initially presented it.

The Quantum Postulate

One crucial premise of Bohr's thinking is what he called the "quantum postulate." By this he means the discontinuous changes in physical quantities (for example energy) at the microscopic scale. When an atom changes state, for example, the atom's energy changes discontinuously from its initial value to its final value, without passing through any intermediate values. All exchanges of energy in the quantum world share this property, essentially due to the fact that Planck's constant is not zero.

Why is the fact of this discontinuous change important? To understand the importance of the quantum postulate, we must look at the contrasting situation in Newtonian classical mechanics. In classical mechanics, we may add energy to or subtract energy from a system controllably and continuously, letting the amount become arbitrarily small if we wish. In fact, this exchange of energy may asymptotically approach zero. It is just this continuity in classical physics that allows us to precisely define the state of the system. The corresponding discontinuity in quantum physics in turn prevents us from so defining the system's state. Because the ability to precisely define states is central to classical determinism, the implication of the quantum postulate is that such determinism is no longer possible.

The foregoing bold assertion requires more proof, so let's further examine Bohr's reasoning. A second crucial premise is that we only know the properties of a physical system by interacting with it. A totally isolated system has no real meaning for us, because it can disclose no information. Whether such a system even exists is an ontological question, but for all epistemological purposes the system might as well not exist, since we can't know anything about it. In more functional language, the object we wish to study must always be considered as part of a larger system that includes the instrument we use to study it. As a purely practical matter, this statement is just a truism of unknown import; in Bohr's quantum interpretive schema, however, the necessity of including the measuring instrument is elevated to an important philosophical principle. Hence, our knowledge is limited, as a matter of principle, to knowledge gained through interactions with the object we wish to know about. But now combine this premise with the quantum postulate: during the necessary interaction, some uncontrollable amount of energy will be exchanged between the object and the measuring instrument, leaving the object in an undefined state. Since both the quantum postulate and the interaction are necessary consequences of the act of knowing, then it is legitimate to question whether the object ever had a well-defined state to start with (after all, such a hypothetical state is in principle outside our ability to know).

We do, of course, gain a great deal of information during the interaction between the object and the measuring instrument. In particular, we can specify the location of the object in space at a well-defined time. This ability is important, because specifying the space and time coordinates is absolutely essential to our conventional notions of how to understand physical reality. Predicting how the location of a particle in space varies with time is, in a way, the fundamental aim of classical Newtonian mechanics. Hence, we highly prize retaining the ability to create a space-time description of physical systems, as we do in quantum theory.

The price we pay to retain this ability, however, is the uncontrollable exchange of dynamical quantities like energy and momentum during the interaction. The reason this price is so crucially important is that conservation of dynamical quantities like energy and momentum constitutes the bedrock physical principle upon which we build the entire edifice of our physical understanding of nature. More particularly, these conservation laws are intimately tied to our usual notions of causality. Due to the essential discontinuity in dynamical quantities introduced by the quantum postulate, we no longer have well-defined causal relationships in our description of the system. But causality has been a central concept in the sciences since antiquity, and the heart of Newtonian dynamics is that the application of a force causes a well-defined and well-understood change in the motion of a particle. To renounce causality would be to give up any hope of doing science at all. Fortunately, we don't need to renounce causality altogether, because a physical system can be prepared in a perfectly well-defined dynamical state. There is no restriction on our knowing the energy and momentum of a system, from which we can calculate its future behavior using our dependable conservation laws. So, a causal description is perfectly possible. The catch is that in order to do so we must renounce our ability to know the system's space and time coordinates.

Complementarity

Hence we have now concluded that a description of a physical system in terms of space and time is possible, and that a description of the same system in terms of cause and effect is possible, but that these two descriptions are mutually exclusive of each other. And yet, both of these descriptions are absolutely necessary if we are to have a complete understanding of the system. The word that Bohr chose to characterize this situation is *complementarity*. We have two *complementary* descriptions, each of which excludes the other and yet both of which are needed for complete understanding. Bohr summarizes the situation thusly: "On one hand, the definition of the state of a physical system, as ordinarily understood, claims the elimination of all external disturbances. But in that case, according to the quantum postulate, any observation will be impossible, and, above all, the concepts of space and time lose their immediate sense. On the other hand, if in order to make observation possible we permit certain interactions with suitable agencies of measurement, not belonging to the system, an unambiguous definition of the state of the system is naturally no longer possible, and there can be no question of causality in the ordinary sense of the word. The very nature of the quantum theory thus forces us to regard the space-time coordination and the claim of causality, the union of which characterizes the classical theories, as complementary but exclusive features of the description, symbolizing the idealization of observation and definition respectively."[79]

The use of complementary descriptions, then, is not merely a convenient choice that we might make or not make as we please. Bohr's argument is that the use of complementary descriptions is quite necessary. It's necessary because it is mandated by the contingent physical fact that the quantum of action (Planck's constant, **h**) is not equal to zero. Due to this fact, we have no choice but to employ the two complementary descriptions (space-time description and causal description). The use of one automatically precludes the use of the other; they are mutually exclusive. And yet, both are needed for a complete description of what there is to know. Moreover, using both (though we can't use them simultaneously) is sufficient to provide a complete description of everything that is physically knowable about the system. Note carefully, however, that the quantum postulate (which is a physical fact) has been combined with the necessity for observation, which is an epistemological assertion of principle. Given these two premises, Bohr concludes that complementarity is a necessary feature of how we understand the world.

In contrast, classical physics involved nothing resembling complementarity. Classical physics employs a unified self-consistent description in which all of the variables (space, time, energy, momentum, etc.) simultaneously have precise and well-defined values (recall the last sentence in the passage quoted from Bohr). In fact, it's this very combination of space-time coordination and causality that is the essence of classical determinism (i.e. the idea that all future events are inevitably and predictably fixed). Newtonian mechanics is deterministic in a way that quantum mechanics is not (since space-time and causal descriptions can't be simultaneously used). But why are we obliged to use complementarity in quantum theory but not in classical theory? The reason is that in classical physics, all quantities are completely continuous; in a sense, Planck's constant *is* equal to zero in classical physics. From a rigorous physical point of view, classical results are merely an approximation, appropriate to the macroscopic world, in which all of the energies are so large that **h** can effectively be ignored (i.e. taken to be zero). In this point of view, quantum physics is a more fundamental theory, which approaches toward the classical results in the appropriate limits. For Bohr, then, the classical world picture, in which the goal was to calculate the deterministic space-time trajectories of particles acted on by causal influences, was merely an outdated framework for understanding nature. He considered complementarity to be a new framework within which to understand nature. Put differently, complementarity is a rational generalization of the old classical picture, a new framework that we can now use to interpret and make sense of the seemingly paradoxical results of quantum theory.

Using Complementarity

An obvious application of the complementarity framework is to the interpretation of Heisenberg's Uncertainty Principle (also known as the Indeterminacy Principle). Historically, Heisenberg derived this result from his matrix formalism at virtually the same time that Bohr devised the complementarity framework. In examining the Uncertainty Principle, we see clearly that its content is entirely consistent with Bohr's complementary descriptions. As the position variable is made more precise (thus allowing a space-time description), the momentum becomes correspondingly less precise (*dis*allowing a causal dynamic description). In an identical fashion, as the time variable is made more precise (thus allowing a space-time description), the energy becomes correspondingly less precise (*dis*allowing a causal dynamic description). And the situation is *vice versa* for both cases, obviously. This is a beautiful example, rigorously mathematical, of the fundamental point made by the complementarity framework. In some ways it is too good of an example, because this has misled some people into confusing complementarity with the reciprocal uncertainty relations, or believing that complementarity is merely verbal camouflage to explain away the Uncertainty Principle results, or to focus all their attention on the simple algebraic result and thereby miss the subtle epistemological point that Bohr was trying to make.

In a more sophisticated analysis, however, the canonically conjugate variables connected by an Uncertainty relation are seen to be specific outcomes, derived mathematically, of a more general statement (namely, the complementarity framework itself) concerning what we may know about the world.

Another important application of complementarity is to the problem of understanding wave-particle duality. To say that an electron is both a wave and a particle is self-contradictory nonsense within the framework of classical physics. (It also violates common sense; note that both classical physics and common sense are grounded in our macroscopic experience of the world.) Within the framework of complementarity, however, the notion that an electron (indeed any matter or radiation) *must be described* as both a wave and as a particle is merely a natural consequence of the more general necessity of employing complementary descriptions of physical systems. Recall that a central tenet of the complementarity idea is that the two descriptions must be mutually exclusive, i.e. the use of one description precludes the simultaneous use of the other. In the case of wave-particle duality, this restriction on our use of concepts translates directly into an analysis of the experimental details. In any particular experimental situation, the object cannot behave both as a wave and as a particle at the same time under identical experimental circumstances; this would indeed be self-contradictory if it could happen.

The standard example is a two-slit interference experiment, in which a beam of light is sent through narrow slits onto a screen where it is detected. The slits constitute two sources of light that then exhibit the phenomenon of wave interference, resulting in alternating bright and dark bands on the screen. Imagine now that we make the beam of light so dim that only one photon at a time hits the screen to be detected. We now observe photons hitting the screen at points where the bright bands will exist, with the probability that a photon will hit a certain point being equal to the eventual brightness at that point. In the dark bands, the probability of a photon hitting is zero. Notice that something truly astonishing has happened (and this is a real experiment that has been performed, by the way). A single photon hits the screen, so the light is composed of particles at the moment of detection. But, in order for the interference pattern to be created on the screen, the light must have been a wave while going through the two slits. The key point here is that, in accordance with the complementarity framework, the corpuscular nature of the light cannot in principle be detected. For example, if we ask "which slit did the photon go through?" then we can only answer by putting a measuring device there at the slit; yet if we do that we will disturb the system and destroy the interference pattern at the screen. Bohr and Heisenberg carefully examined many such experimental arrangements and concluded that none could be devised that could violate the mutual exclusivity demanded by complementarity.

The point being made here is centrally important. Why is it not a contradiction to say that the light is a wave in the slits and a particle at the screen? Because we can't *observe* the light being a wave in the slits. This example reminds us of Bohr's emphasis on the role of observation (the second crucial premise that I mentioned earlier). We can only know about an object what we can find out by interacting with it. If we can't observe an object being both a wave and particle in the same act of measurement (and of course we can't), then there is no contradiction. For Bohr, our interpretive problems always arise from an improper use of language and concepts that were formulated within a classical physics framework and invalidly extrapolated to a microphysical realm wherein the concepts simply don't apply. The complementarity framework was devised expressly for the purpose of fixing such problems by providing a more careful and appropriate way to use language and concepts. Put differently, we might say that waves and particles are mechanical pictures grounded in our macroscopic experience (as refined by classical physics), and that trying to use these pictures to explain microscopic phenomena beyond our sense perceptions is inappropriate. This inappropriateness leads to the impossibility of creating a single unified and consistent picture, but we can rescue our concepts by invoking complementarity as a guide to their proper use.

These considerations give rise to two related questions. One question is: If the concepts and pictures of classical physics are no good, why continue to use them at all? The second question has to do with a contention of complementarity that we've ignored so far in our examples, namely that the complementary descriptions are exhaustive (i.e. that between them they tell us all that we can know). On what basis can we claim that our complementary descriptions exhaust all the possibilities of what we can know about a system?

The Use of Classical Concepts

This brings us to Bohr's third crucial premise of the complementarity framework: in order to describe physical phenomena, we must employ the language and concepts of classical physics and the macroscopic world. This idea always remained the vaguest and most problematic part of Bohr's thinking. In one sense, he meant that our experimental observations ultimately consist of sensory perceptions at a macroscopic scale. The measurement apparatus is always classical, in that sense, even if its operation is based on quantum physics (as most instruments today are). In a related sense, we live in a classical world which shapes our language and our visual imagination; thus we must use these to communicate with each other. But he also seemed to imply that the more abstract concepts of classical physics, like energy and angular momentum, are also indispensable (perhaps because these are only refined versions of our more primitive macroscopic perceptions). In any event, the indispensability of using classical concepts appeared to be self-evidently correct to Bohr. He often stated that it is true, but did not offer very much argumentation to demonstrate this. After all, how can we imagine that which is totally beyond any possible experience we may have?

If we grant that this premise is correct, and that we are forced to employ classical concepts to describe microscopic objects, then the complementarity framework follows naturally from the rest of the argument. The quantum postulate introduces an essential discontinuity into our interactions with objects, and such interactions are an essential requirement to know anything about the objects. Space-time descriptions and causal descriptions then of necessity become complementary, and any classical picture that unites such descriptions must perforce be employed also in a complementary way. But is it possible to circumvent complementarity by devising and using a set of new and non-classical concepts that somehow still allow a unified description under all circumstances? Bohr answered that this is not possible, that we cannot transcend our "forms of perception" as he sometimes called them. But other physicists disagreed with this judgment, and advocated a search for new theories along these lines, generally called by the generic name of "hidden variables" theories. If we can find the right set of hidden variables to use, we can do away with complementary descriptions. The motivation to engage in such a program will soon be apparent.

Criticisms of and Alternatives to Complementarity

From the beginning, there was opposition to Bohr's conclusions. Some physicists, preferring equations to philosophy, were merely indifferent. But among the more philosophically inclined physicists, there was also some active antagonism, most notably that of Einstein (which has become legendary). There are two highly important reasons underlying the rejection of complementarity. One reason was the assumption that a proper theory should be deterministic. In other words, critics disagreed with Bohr's contention that we must revise our notion of what we ask from a theory, and instead demanded a theory that matched the ideals of classical physics. In this view, a hidden variables theory would be deterministic (in the new variables) and probability becomes an artifact of using the inappropriate classical variables. A second reason, closely related to the first, is the problematic role of the observer in the complementarity framework. Quantum mechanics becomes less a theory of what *is* and more a theory of what *we know*. This change in emphasis was thoughtful and intentional on Bohr's part, but was (and still is) highly controversial.

The relationship between these two issues (determinism and the role of the observer) is thrown into sharp relief by an examination of the infamous "measurement problem" in quantum theory. The problem is this: We can only calculate the probability of finding a specific value for some physical quantity. We then measure this quantity experimentally, and it now has a specific value with certainty. How did our act of measurement alter the state of a physical system from one having a probable value into one with a definite value? Bohr's heavy emphasis on the need to always include the measurement apparatus along with the studied object as parts of a larger system ultimately invites this question, and neither quantum theory nor the complementarity interpretation can answer it. Beyond its importance as a technical question in quantum physics, this issue has deep philosophical implications, because it implicitly questions the very existence of an objective reality free of observers.

Such problems reach a crisis when we try to analyze the EPR paradox. In order to demonstrate the flaw in the EPR argument, Bohr needed to distinguish more carefully between the physical system as an independently existing object with properties (which we either know or don't know) and the physical system as a "phenomenal" object possessing properties only insofar as an interaction measures them. Put more bluntly, Bohr now had to deny the existence of any intrinsic properties of a physical system, independent of the conditions of observation. While the logic of this position allowed him to refute the EPR claims, the cost was an even deeper intrusion of the observer into the conception of physical reality. An objective view of reality can be retained under these conditions, but not one that is satisfactory to the critics of complementarity.

The main rival of the complementarity interpretation involved devising a set of hidden variables, in terms of which the resulting theory would be deterministic and observer-free. Such a program was not pursued for many years because von Neumann had published a mathematical proof that such theories were not possible. Decades later, a flaw was found in his proof indicating that hidden variables theories might be possible if these variables are allowed to be non-local. Meanwhile, Bell's analysis of the EPR paradox and the subsequent experimental work also showed that non-local effects must be at work. The situation is consistent with either the complementarity view or some form of hidden variables theory. To date, no successful theory of this sort has been produced, but some investigators are continuing to work in the area.

At present, the so-called Copenhagen interpretation, including complementarity, is still the mainstream orthodox interpretation of quantum mechanics. In addition to the hidden variables theories, a variety of other alternative inter-pretations such as statistical ensembles, decoherence, splitting universes, quantum logic, consistent histories, and so on have been proposed. The major goal of many of these attempts is to recover a theory that is deterministic and that does not involve the observer. Motivated by this goal, work continues. Bohr's strategy was to follow the lead of nature and accordingly revise what we ask for from our theories. In doing so, he believed that an important epistemological lesson had been learned, a lesson that also carried implications for other fields beyond quantum theory.

Extension of Complementarity to Other Fields

Although Bohr devised complementarity in response to the factual content of quantum mechanical investigations, there is good reason to believe that he was influenced in his thinking by general epistemological considerations (learned through his contact with Hoffding), and he did not confine the potential applications of complementarity to the understanding of microphysical reality. Rather, he stated on a variety of occasions that he thought these "epistemological lessons" uncovered by the attempt to understand quantum theory would have wide application in many fields of knowledge. Although subsequent development in these fields has not followed up on Bohr's suggestions, it's worth a brief look at some of the examples he has given and why he thought complementarity would be of value in these cases.

In psychology, Bohr thought that the complementarity framework would illuminate the problem of free will and determinism. Just as a microscopic system must be described in the context of the means to measure its properties, so our own actions must be described in terms of the actions themselves (objects) while including the means of observation (the knowing subject). The actions may be described in isolation, analogous to the space-time descriptions of physics, and within such a description the action may be explainable as being determined by prior circumstances. This description, however, must be only a partial view of reality; for a complete view, we also need the description that includes the subject (or observer), analogous to the inclusion of interactions in physics. This leads to a complementary explanation in terms of the action resulting from an instance of conscious will. These two explanations seem to be contradictory, but this is only because insufficient attention was paid to the circumstances underlying them. The key point is the necessity for both descriptions (though each excludes the other). Hence, the complete explanation of the action must include both free will and determinism as complementary pictures of how the action comes about. Both are true, neither is complete, and each follows from its own arrangement of the subject/object relationship.

Biology was another field in which Bohr advocated the use of complementarity. The issue here is whether a purely mechanistic description (using physics and chemistry) is capable of exhaustively explaining all biological phenomena. Bohr's argument is that this is not possible, because the conditions of observation needed to study living organisms are incompatible with the conditions needed by definition for a mechanistic description. More specifically, a living organism cannot be treated as an isolated system. All living things must interact with their environments in order to remain alive, exchanging material and energy continuously. Thus a mechanistic description requires as a matter of principle conditions of observation antithetical to life. Following this line, he concludes that in addition to the mechanistic description (which is indeed necessary), we must also employ a complementary teleological description based on purposes and functions. As usual, the two kinds of explanation are mutually exclusive but not contradictory since each can only be used under its own characteristic circumstances. And both kinds of explanation are needed to achieve a complete understanding of the organism.

These two examples were the ones that Bohr developed most fully. He also outlined, in a brief and sketchy way, how complementarity might be applied to cultural anthropology and to ethics. His major point in all of these discussions centered on the appropriate use of concepts, and on the need under all circumstances to account for the conditions of observation in our acquisition/formation of knowledge. Whenever such conditions do not allow the unambiguous isolation of the object of knowledge, it becomes necessary to use complementary descriptions of this object in order to achieve a complete understanding of it. This need for complementarity in our interpretation of physical phenomena is the epistemological lesson that quantum mechanics teaches us, and Bohr believed that this lesson is widely applicable in our attempts to understand the world.

8. Revisioning Complementarity

We now come to our main task. In its most general meaning, complementarity resolves antinomies by allowing valid truth claims for more than a single member of a set of contraries. We are concerned with a pair of contraries: Nature is mundane; nature is sacred. Nature is mundane, and a valid understanding of nature requires only the propositions of scientific materialism. "...everything that exists or occurs [is] conditioned in its existence or occurrence by causal factors within one all-encompassing system of nature..."[80] Nature is sacred, and a valid understanding of nature implies deeper and broader dimensions than materialism allows. "...the order of nature is related to an order 'beyond' itself, to what we might call 'spiritual principles.'"[81] Can we accept both of these statements as valid? My claim is that we can do so, and that we can do so not by evasive verbal games but by a meaningful application of the framework of complementarity to the issues. In fact, my claim is that we have no choice but to accept the validity of both statements.

I will not, however, employ complementarity as Bohr developed it. We are dealing here with a much more extensive set of issues than those confined to the use of concepts within objective sciences, and Bohr's development (including his anticipated broader applications) was solely confined to these sciences. His detailed development was in fact confined only to problems in atomic physics. My program here is to look closely at Bohr's work and to base upon this work a set of generalizations and analogies in order to develop a generalized version of complementarity suitable for analysis of our problem.

Extended Analogy and Generalization

A crucial premise in Bohr's work is the undivided wholeness of the observed system and the means used to observe it (experimental apparatus). In essence, an isolated system, not being observed, has no meaning (or at best, an ambiguous meaning). So the presence of an observer is required to impart meaning to the observed phenomenon, but the "observer" in Bohr's interpretation is merely some instrument that records (non-erasably) the result of an experiment; no conscious understanding need be involved. (This point has caused misunderstanding and controversy. Although some interpreters of quantum mechanics have incorporated conscious observers into measurements, Bohr did not). The necessity of including the means of observation in order to give meaning to a phenomenon is at the heart of Bohr's thinking, though, and is integral to the validity of complementarity. Thus, I will analogously make the observer an integral part of the process by which we apprehend nature. In the present generalization, however, this observer *does* need to be an actual conscious human, a knowing subject.

The need for a conscious subject as the observer in our present context seems fairly obvious, given the complementary conceptualizations of nature that we are considering. The justification for including this subject, however, and the limits on the validity of any conclusions we may draw based on the inclusion, are not so obvious. These issues will be further explored in due course. For now we'll simply note that the issues are quite general, and form an important part of the so-called critical philosophy. Going back at least to Immanuel Kant, the subject/object relationship has been an important epistemological problem, to which Bohr's mentor Hoffding made important contributions. Bohr was aware of these implications in his work. "...complementarity is suited to characterize the situation, which bears a deep-going analogy to the general difficulty in the formation of human ideas, inherent in the distinction between subject and object."[82] The ramifications of the necessity to include a knowing subject are central to our analysis, but need to be postponed until all the terms of our generalized complementarity framework have been introduced.

In atomic physics, the inseparability of the observer from the object of knowledge carried a drastic implication: knowledge of the object is no longer independent of the conditions of observation. This idea is at the heart of complementarity. Because knowledge of the object depends on the conditions of observation, we must carefully specify these conditions in order to have any meaningful knowledge at all. In atomic physics, of course, the specifications concern experimental arrangements and the desired knowledge is objective knowledge about the system, which Bohr refers to as "unambiguously communicable information"[83]. For our problem, knowledge concerning nature will not necessarily be objective. But, the crucial importance of examining and specifying the conditions of observation is once again implied by the inclusion of an observer, just as in Bohr's interpretation of quantum mechanics. How the specification is performed and its inclusive scope are considerably different for the case of a knowing subject, extending well beyond the mere description of an experimental arrangement. The kinds of questions being asked; the state of consciousness of the observer; the modes of communication possible and those employed; the role of multiple observers and/or technology used in observation; the effects of culture and history, of time, place, and intention; all these things must be taken into account in order to understand the meaning of any knowledge we may have of nature. To examine more carefully these kinds of conditions in a variety of particular cases, and attempt to draw appropriate conclusions, will be a recurring theme in this work. The present point is simply to note that it needs to be done, and if our analogy holds true then a proper specification of the conditions under which knowledge of nature is acquired results in the complementarity framework being free of logical contradictions. The conditions under which nature is found to be sacred are not those under which it is mundane; both sets of conditions are valid and hence necessary for a complete view; and the sets of conditions (and knowledge derived from them) must be somehow correlated since they describe, at root, the same world.

We come now to a major premise in Bohr's work for which there is *no* analogy in our generalization, namely the need to employ classical concepts. The reason for this breakdown in the analogical treatment is important and illuminating. The classical concepts (like space-time descriptions, causality, energy, and momentum) needed to be employed because they were intrinsic parts of the integrated worldview evolving naturally out of our macroscopic perceptions. These concepts are, in this sense, just a natural extension of ordinary language, the very language in which we conceptualize our world to start with. This world is an integrated world (by and large, at least) which hangs together, and the refined classical physics picture based on it very definitely hangs together. Space-time coordination and causality work together seamlessly to produce a complete, coherent, and self-contained worldview. It was this worldview that was shattered by the discovery of the quantum ($\mathbf{h} \neq 0$) and the resulting need for complementarity. In stark contrast, no such coherent and self-contained view of nature in the broadest sense (simultaneously religious, philosophical, scientific, rational, mystical, and empirical) has ever existed. Therefore, no analogy to the limitations on the classical picture imposed by quantum theory and complementarity can exist, because no analogy to the classical picture itself exists. Put a little differently, it's as if we imagined a "wave theory school" and an opposing "particle theory school" as contending factions trying unsuccessfully to achieve a hegemonic picture of macroscopic physical reality, with the impasse finally broken by the introduction of complementarity. Of course, this never happened because both of these (waves, particles) fit coherently into a single overarching picture. Why? Because at the macroscopic level, Planck's constant is so small that we can effectively take it to be zero. If we had lived in a universe where **h** was not small by macroscopic standards, we might imagine the ensuing intellectual strife in the development of physics. In view of the analogy we're building, the apprehension of nature (as mundane or as sacred) has been developed in just such a universe: whatever plays the role of a "quantum of action" in this analogy is not small, and we do indeed suffer from controversy on this question. Hence arises the need for complementarity.

But there is an important part of Bohr's premise that still survives here. Although no analogy to the "classical concepts" exists in the sense just discussed, those concepts that we are in fact using and that collectively make up the "mundane" and "sacred" rubrics (as we explored them earlier) are concepts that are formed subject to just those kinds of constraints and conditions stipulated by Bohr. Our concepts must ultimately be limited by our experiences, our language, and the capabilities of our minds. This limitation can't be transcended and we need to use the concepts we have available to formulate our view of nature, even if the analogical "**h**" has turned out to be large for the world we inhabit.

The question of what actually does play the role of **h** in this analogy is difficult to answer. In Bohr's framework, it is the existence of **h** that entails the necessity of joining the observer and observed into an undifferentiated wholeness. In more general epistemological terms, the necessity to include the knowing subject in any consideration of the apprehension of a phenomenon has long been known. For example, Kant wrote that "...if the subject, or even only the subjective constitution of the senses in general, be removed, the whole constitution and all the relations of objects in space and time, nay space and time themselves, would vanish."[84]   There is certainly no question about the need for a perceiver in order for there to be a perception, but the influence of this knower on phenomena (and whether anything lies behind the phenomena) has been highly controversial, involving a thick tangle of philosophical issues. Slicing through this tangle is the fundamental question: does how we know the world make any difference to how the world is? Imparting importance to this question is the fundamental fact: the only world we have is the world that we know. This fact is, perhaps, the best answer possible for the question of what plays the role of **h** in our analogy. But the answer to the question of what difference this makes to the world itself is complicated and will be deferred to the following chapter.

Despite admittedly pushing several tough issues temporarily into the background, we have achieved our initial goal of presenting the main points of a generalized complementarity framework capable of addressing the sacred/mundane antinomy in the apprehension of nature. To summarize these main points: There is an inherent inseparability between the knowing subject and the known world in our apprehension of nature. This is so because we only know the world through our experience of it. This experience is conditioned (in some sense at least) by our minds (broadly considered, including language, culture, neurophysiology, and so on), limiting our available concepts. Due to these limitations, our understanding of nature is only meaningful when the concepts that we use and the conditions under which we use them are carefully examined. The results of such a careful examination yield clusters of concepts (which we can conveniently label "nature as sacred" and "nature as mundane") that appear to be contrary descriptions but that in fact are indicative of differing and mutually non-overlapping conditions of knowing; both descriptions are necessary in order to have an exhaustive description of nature itself.

Examples and Applications

Let's look at a few specific cases to get a sense of how these ideas might apply to the world. The issues become more pointed when questions of life and consciousness are involved, so we'll start with a deep and powerful instance, namely the manner in which the body of a dead person is considered. "Matter is indeed infinitely and incredibly refined. To any one who has ever looked upon the face of a dead child or parent the mere fact that matter *could* have taken for a time that precious form, ought to make matter sacred ever after. It makes no difference what the *principle* of life may be, material or immaterial, matter at any rate co-operates, lends itself to all life's purposes. That beloved incarnation was among matter's possibilities."[85] These stunning words of William James point directly to the sacred dimension inherent in the world. Within its own context, I find it hard to imagine contradicting this statement. And yet, the believer in a mundane world has plenty of valid arguments to mount: These atoms, making up the body, are no different than any other atoms. The chemical bonds are identical, and subsequent to the body's decay these atoms may well drift into inorganic and less-sacred-seeming forms. Aren't we simply mistaking our own feelings for reality here?

In accordance with the analysis presented earlier in

the chapter, we need to look closely at the "conditions of observation" in each case to judge the validity of the two views. Specifying these conditions first entails specifying the kind of knowledge we require. Knowledge concerning the person who is now dead, knowledge of this person's life and individuality, is of a different order than knowledge of the dead body's anatomy, of its chemical composition, of its imminent decay. The first kind of knowledge may confer a sacred quality to the dead body, the second kind of knowledge may not. Also, whether there is a strong existential relationship between the "observer" and the dead body can make a profound difference here. That this is so brings us to one of the major issues at stake here: how can the presence or absence of such a relationship make any possible difference to the status of something in the world? Doesn't the "object" have its own independent existence with its own properties? It appears that we are drifting toward a subjectivistic, idealist, and solipsistic world-view, to which the partisan of mundane existence will object; the body is truly no more than its chemical composition and unbinding atoms, and this sacred quality you perceive is no more than your own emotions within your own mind, not a quality inhering in the thing itself. But this is the very epistemological point that is the crux of my entire argument, because any qualities inhering in the thing itself are *a priori* unavailable to us. We can only know what is available to our senses, our reason, our minds. The body construed as a collection of elements temporarily found in a special configuration is a construction within our minds, no more and no less than the body as the sacred remains of a loved one. I am not promoting solipsistic idealism here. The body is real and possesses qualities, but the only qualities it can posses for us are those it possesses by virtue of its interaction with us. We do not need to deny the mundane dimensions of the properties of the body in order to assert its sacred dimensions. Both are true, hence neither is exhaustive, and no contradiction exists between them (consistent with our examination of the "conditions of observation"). Our differing apprehensions of the meaning of the dead body are complementary.

I admit that I've left out one important point, the trump card of the believer in a mundane world. The view of the body as a collection of chemical elements is an objective view, i.e. multiple observers can all agree on this same chemical composition. In contrast, the sacred qualities of the body may or may not be shared by different persons, and always has in some sense an intensely personal aspect. The objectivity of the mundane view is taken by its supporters to prove its superiority and even its truth. A formal treatment of this issue must await the next chapter, but even here I can point out that objectivity entails losses in addition to its acknowledged virtues, and that our attitudes toward objectivity are part of our cultural history. In an extended discussion of the role of the body (both dead and alive) during the profound cultural shifts accompanying modernism, Romanyshyn expands on these points. "Within the linear, and homogeneous, space of explanation, within that grid where all space has become equal and the same, the heterogeneous pantomimic body has no place. It is a body, therefore, which we no longer need, a body which has become an obstacle; a body for which there is no place is a body ready to be abandoned. It is also, on the other side of this abandonment, a body ready to be reinvented. The corpse is the most visible image of the abandoned body. It is what the human body becomes in our increasing distance from it. It is what the pantomimic, e-motional body becomes for a spectator self behind a window with a heady vision fixed upon an infinite horizon."[86]

Our second example is closer to what one usually associates with "nature" as the object in question. Many places or natural formations have some sacred meaning within a cultural context, and these are not limited to animistic projections and primitive cultures. Consider these comments from a sophisticated Japanese practitioner of Zen Buddhism: "The Japanese love of Nature, I often think, owes much to the presence of Mount Fuji in the middle part of the main island of Japan. Whenever I pass by the foot of the mountain as a passenger on the Tokaido railway line, I never fail to have a good view of it, weather permitting, and to admire its beautiful formation, always covered with spotless snow and 'rising skyward like a white upturned folding fan,' as it was once described by a poet [Ishikawa Jozan] of the Tokugawa period. The feeling it awakens does not seem to be all aesthetic in the line of the artistically beautiful. There is something about it spiritually pure and enhancing. [...] Fuji is now thoroughly identified with Japan. Whenever Japan is talked or written about, Fuji is inevitably mentioned. Justifiably so, because even the Land of the Rising Sun would surely lose much of her beauty if the sacred mountain were erased from the map. [...] In the beginning, probably, the Japanese were naively attracted to the beautiful which they saw about them; it is possible that they regarded all things in Nature as uniformly animated with life, after the manner of primitive people who look upon even nonsentient things from their animistic point of view. But as they cultivated themselves in the Zen teaching, their aesthetic and religious sensitiveness was further nourished. And this nourishment came in the form of an exalted moral discipline and a highly spiritual intuition. That is to say, the snow-crowned peak of Fuji is now seen as rising from the background of Emptyness..."[87] But Mount Fuji is merely a geological formation, resulting from volcanic forces understood within the context of plate tectonic theory. The snow capping the mountain is a meteorological phenomenon, a vapor-solid phase transition like any other snow. Do these seemingly undeniable facts negate the sacred quality so eloquently described by Suzuki?

Those words were written by a Zen scholar, so Mount Fuji

is observed through the eyes (and mind) of a Zen disciple. In Zen *satori* is found a meaning that "is not something added from the outside. It is in being itself, in becoming itself, in living itself."[88] This direct and unfiltered experience of Mount Fuji is different from the experience of the mountain that results from a normal state of consciousness. The resulting meanings and forms of knowledge from the two states are correspondingly different. A direct experience and intuitive knowing cannot yield knowledge of geological history or thermodynamic transformation. Conversely, an analytical consideration of temperature, composition, humidity, and other quantitative data cannot yield the realization of sublime beauty. These two vastly differing states of consciousness thus constitute the "conditions of observation" for the complementary views of nature in this example. Having identified them, we easily see why the two views don't directly contradict each other, since the two states of consciousness won't simultaneously coexist (direct experience and reflective analysis must surely exclude each other). It still remains to demonstrate that each of these realizations of Mount Fuji is valid and correct. Toward that end, I would point out that each view has the support of an entire cultural infrastructure and an extensive literature over many centuries. Proponents of either view can and do make strong claims for its validity. I simply endorse both sets of claims.

The last example is also from nature, but not a culturally important object with iconic status for a large group of people. Instead, let's consider a humble piece of crystalline mineral, such as quartz, citrine, or garnet. Crystals are aesthetically beautiful in both sacred and mundane perspectives. Even as mundane objects, crystals hold a tremendous amount of interest. The atoms making up a crystal form a regular periodic array (called a lattice) that has many fascinating mathematical properties. The crystal lattice has a variety of important symmetries (described by the branch of mathematics called Group Theory), which manifest themselves macroscopically in the exquisite facetted shapes of the crystal. The symmetries of a crystal are also responsible for our ability to understand and calculate its physical properties. As just one characteristic and important example, the symmetries of the crystal lattice produce the so-called "band structure" of its electron states, which is the ultimate basis for all microelectronics technology. For many people, however, the interest of crystals stems not from their mathematical structure but from the sense of mystery and power they possess. Crystals, for such people, are sacred objects. Shamans employed them as such millennia ago, and still today there are people who do so. Is there any validity in such a notion, and can it be consistent with our highly developed scientific understanding of crystalline properties?

I think that upon investigation we'll find the answer is yes to both questions, but I'll start by pointing out that for me the mundane view is clearly unquestionable. I have personally investigated experimentally the properties of crystals for almost three decades, and never cease to find their physical and mathematical properties fascinating. What are the "conditions of observation" that characterize this endeavor? The key condition here is that we restrict our range of interest to consider only properties that are numerically quantifiable (e.g. positions of atoms in the lattice) and hence measurable. This seems on its face like a hugely restrictive condition, and it is indeed extraordinary that such a fruitful and profound representation emerges from such a seemingly barren starting point. For all its virtues, however, this view does exclude any kind of sacred quality to the crystal. Nor does such a sacred quality contradict anything in the mundane view either; there is simply no room for it. When I "look" at the crystal under the mundane conditions, I can "see" only what I allow. We need now to consider whether there is some other alternative "conditions of observation" that we can define so as to make room for a complementary view.

Such an alternative is offered by phenomenology, a philosophical methodology that "puts essences back into existence [... and ...] tries to give a direct description of our experience as it is..."[89] When I "look" at the crystal under conditions set by a phenomenological approach I may "see" an entirely different aspect of the existence of this crystal. "A real being has more materiality, and more significance, than I intend to perceive; when I perceive the real, I must adjust my general intentions to the apparitions which emerge from yonder depth [...] A real being shows me unsuspected meaning in every perception aimed toward its density; from yonder depth emerges significance beyond the meaning which I had in mind. Real materiality progressively releases hidden meaning during our mental exploration."[90] If we apprehend a crystal employing "conditions of observation" set by a phenomenological approach, we may apprehend emerging from its "yonder depth" a significance that we can justifiably label sacred, a significance not found in our previous encounter with the crystal. Now, our crystal is no less sacred when we encounter it within the context of our mundane world; its sacred quality was merely less apparent under the conditions we encountered it then. Nor are its mathematical and physical properties any less a part of the total reality of the crystal when it's encountered as a sacred object; once again, these properties are merely less apparent then. These two different apprehensions of the crystal are complementary: they are mutually exclusive, both are needed for a complete understanding of the crystal, and they do not contradict each other because the conditions of observation needed for each do not overlap. These conditions, as described here, are quantitative and analytical on the one hand and a phenomenological methodology on the other hand. Yet, it may be misleading to say that these complementary views don't overlap, because after all it is the same crystal. This crystal combines within its totality both of these views of itself, and we who encounter it ultimately do the same.

Brief Afterword

The alert reader will have noted that each example was carefully arranged to avoid possible genuine contradictions between the complementary views. Complementarity claims that this will occur naturally, but it's not hard to see ways in which our examples might be made more complicated. Include a contention that the personality survives death in example 1, or ascribe some particular healing power to the crystal in example 3, and suddenly there is a murkier overlap between the complementary views along with the possibility that contentions made within one view might contradict contentions legitimately made within the other view. If my argument is valid, then a more careful analysis of the conditions of observation (which is to say, the grounds justifying knowledge) for each view should clear up the problem while also providing more insight and understanding of the issues pertaining to the case at hand. I've avoided this process in the present examples because they are only intended to illustrate the idea behind, and the use of, our newly generalized complementarity framework. The extended examples in the last three chapters will confront such complications, so these will represent a more rigorous test of the method while also presenting more of the subtleties that might arise. Meanwhile, we'll next complete another undone task by justifying the contention that the complementary views are equally valid, and not merely competing alternatives of which one is simply wrong.

9. Epistemological Questions

I have so far presented arguments for the proposition that humans need to experience nature both as sacred and as mundane, that an inherently religious dimension of reality is on the one hand necessary and on the other hand does not contradict scientific materialism (itself also necessary) within the context of complementarity. If one accepts these arguments as persuasive, then two important and interrelated issues still remain to be resolved: The first question is whether these two complementary faces of reality are attributes of reality itself, or whether they are merely descriptions that we ourselves impose on reality. In other words, is complementarity saying something about the world or something about our psyche? The second question concerns the role and interpretation of objectivity within the context of our complementary views of nature. Should objectivity claim a privileged status in our evaluations of the complementary views, and if so isn't there a bias in favor of scientific materialism? The second question clearly relates back to the first, because such a bias would suggest that the materialistic view stems from a genuine attribute of reality while the sacred qualities of nature are merely our own subjective projections. The questions are fundamental and difficult.

We might ask, though, whether the questions are truly important (or at least in what sense they are important). As

William James puts it, "what concrete difference will its being true make in any one's actual life? How will the truth be realized? What experiences will be different from those which would obtain if the belief were false?"[91] Based on these criteria, there are several grounds on which the question of whether complementary understandings of the world are genuine assertions about the world's properties, as opposed to our internal mental categories, is an important question. Making this point, Huston Smith notes colorfully "but how much comfort can we draw from [feelings of awe and reverence] when the awe nature awakens in human beings is, like all emotions, no more than a Post-it note, so to speak, affixed to a nature that is unaware of being thus bedecked."[92] But the most undebatable pragmatic argument for importance is that the truth of an assertion will change how we act, and consideration of the effect of nature's sacredness on our response to the worldwide environmental crisis settles this question quickly (a later chapter analyzes this issue in more detail). Beyond our actions, however, the question remains existentially important. "The very question whether our knowledge is limited to empirical states of affairs implies the question of the transcendent ground of intelligibility; and by asking it one has already pressed beyond the empirical and is open to the existence of transcendent intelligibility. To deny the validity of the question of transcendence is to mutilate human consciousness...."[93]

Fundamental Presuppositions

I start from the premise that we are all limited in our possible understanding of reality by the deeply embedded fundamental presuppositions that we adopt. No discussion of the unification of forces in the early life of the universe could possibly be meaningful to Cardinal Newman, for example, who took scientific work to be the empirical study of eternally existing laws of nature. No discussion of the attributes of angelic beings could make sense to someone who lives in a world from which the existence of such beings is excluded. Debating answers is not an option when the very questions themselves are inherently incoherent. The language of "horizons" is useful here. In the same way that the horizon of a visual field sets a limit to what can be seen in the world, a metaphysical horizon sets a limit to what can be meaningfully comprehended; in both cases, the standpoint of the seer determines the location of the horizon. "Literally, a horizon is a maximum field of vision from a determinate standpoint. In a generalized sense, a horizon is specified by two poles, one objective and the other subjective, with each pole conditioning the other [...] Thus, the horizon of Pure Reason is specified when one states that its objective pole is possible being as determined by relations of possibility and necessity obtaining between concepts, and that its subjective pole is logical thinking as determining what can be and what must be. Similarly, in the horizon of critical idealism, the objective pole is the world of experience as appearance, and the subjective pole is the set of *a priori* conditions of the possibility of such a world. Again, in the horizon of the expert, the objective pole is his restricted domain as attained by accepted scientific methods, and the subjective pole is the expert practicing those methods; but in the horizon of the wise man, the philosopher of the Aristotelian tradition, the objective pole is an unrestricted domain, and the subjective pole is the philosopher practicing transcendental method..."[94] We ask for a demonstration of the fact that nature is *truly* sacred, outside of any beliefs we may hold. But can any such demonstration be convincing to a person whose horizon automatically excludes the meaningfulness of the statement to be demonstrated? I suspect not. What is then required is to change the location of this person's subjective pole so as to bring the possibility within a newly enlarged horizon.

I might add that the reverse is equally true: a person whose horizon excludes the possibility of scientific materialism cannot be convinced of its validity by any sort of argument, only by an enlargement of their horizon. No array of empirical facts or deductive logic will necessarily suffice to alter a person's basic horizon, because the horizon itself sets the limits for what will count as a valid fact or deduction; only within a given horizon can controversies be settled in this fashion. The kind of argument we must attempt here, then, is at the deepest levels of epistemology and ontology. The argument is for the ontological validity of complementarity and hence of both the sacred and the mundane dimensions of the world. Although the success of this endeavor warrants both the sacred and the mundane qualities, I will focus my attention mostly onto the sacred, because the prevailing worldview of the presently dominant global intellectual and political culture clearly favors the mundane.

In this vein, let's consider several statements in which the sacred and spiritual dimensions of reality are simply taken for granted. Discussing what he means by Imagination, Corbin writes: "In this context of agnosticism the Godhead and all forms of divinity are said to be creations of the imagination, hence unreal. What can prayer to such a Godhead be but a despairing delusion? I believe that we can measure at a glance the enormity of the gulf between this purely negative notion of the Imagination and the notion of which we shall be speaking [...] we answer as though taking up the challenge: well, precisely because this Godhead is a Godhead, it is real and exists, and that is why the Prayer addressed to it has meaning."[95] The topic of that work is a 12th century Sufi mystic (Ibn Arabi). Turning to a different tradition, the Vedanta, we find this assertion: "Our metaphysical analysis also shows that the self which is consciousness is being (*sat*) and bliss (*ananda*) as well. It is being because its non-existence cannot even be conceived.

Without its being, no experience is possible [...] Existence, consciousness, and bliss are neither parts nor attributes of the Self; they are its essential nature [...] They are not properties or attributes, because the function of properties is to limit or condition that to which they belong. The Self is unconditioned and non-dual [...] Wisdom lies in transcending appearances and realizing the true Self, which is not the ego as opposed to non-ego, but the reality of both."[96]   Statements like these are typical of an extensive literature based on attempts to communicate the insights discovered within the context of mystical experience.   In the introductory comments to a large collection of such statements, it's written that "in every age there have been some men and women who chose to fulfill the conditions upon which alone, as a matter of brute empirical fact, such immediate knowledge can be had; and of these a few have left accounts of the Reality they were thus enabled to apprehend...."[97]   These kinds of first-hand experiences are not within the horizon of every person (or even many persons), yet they are found consistently in virtually every culture and religion over three millennia.   They certainly constitute important evidence that deserves to be taken very seriously, yet taken alone these accounts will not necessarily convince a person for whom such experiences are alien (after all, a delusion shared by many people is still a delusion).   To proceed further, we need to look more closely at the epistemological foundations of our knowledge.

The Ground of Knowledge

According to Kant, our sources of knowledge can be divided into the content upon which knowledge is based, supplied by the outside world, and the forms that act upon this content, forms that are inherently imposed by the human subjective knower. These forms include the forms of perception (such as space and time) and the forms of understanding (such as causality and quantity). Of the content itself we can know nothing, because all that we know is by virtue of the forms acting on the content; hence Kant postulates an absolute reality consisting of the famous things-in-themselves (noumena) that are forever unknowable. Our known world consists instead of phenomena. The critical philosophy of Kant is thus a nuanced balancing act between the central role of the subjective knower and the objectivity imposed by the existence of an absolute externally real world. Bohr's philosophical mentor, Harald Hoffding, made deeply important contributions to the critical philosophy initiated by Kant, examining both the role of cognition that undergirds the theory of knowledge and the implications of this theory of knowledge for understanding the ontological foundations of Being.

A key point for Hoffding is the role of continuity in the formation of understanding and knowledge. Continuity refers here both to psychological conditions (memory, experience, personality) and to characteristics of the phenomenal world (space, time, causality). For Hoffding, the idea of continuity and discontinuity formed the basis for a unified treatment of cognition, knowledge, and metaphysics. Our very ability to formulate any coherent sense of reality at all depends, according to Hoffding, on the existence of these continuities, and he states that Kant "was not mistaken in declaring that the *demand for unity and continuity* lies at the bottom of all the forms through which we win or expect to win understanding. He himself has shown that all his categories can be traced back to the concept of continuity [...] Logical principles, the principle of causality, and the fundamental doctrines of natural science, all hinge on this concept..."[98] In the broadest sense, this concept of continuity emerges from the fact that nothing is ever known in isolation but only in its relationships to all else. Those cognitional elements characterized by continuity then find commensurate aspects of Being that also possess such qualities of continuity, and from this process issues forth valid understanding. "What appears as an hypothesis from the purely empirical view, becomes, epistemologically considered, a principle, a regulative thought, under whose leadership consciousness may satisfy in the empirical world its demand for continuity and union [...] The idea of a working hypothesis points in two directions: on the one hand, as already demonstrated, back to the nature of the thinking consciousness [...] on the other, to the reality to which the phenomena to be understood belong. A tool must be adapted both to the hand that is to use it and to the object to be worked on."[99] Thus, from the impressions, dreams, fragments, and sensations passing through our minds, the principle of continuity is used to construct and validate reality.

But there is discontinuity as well as continuity. "Even after fruitful principles or working hypotheses have been attained, will Being be completely rendered by them? Or will there always remain an irrational relation *between the principles which may compose our consciousness and the Being itself from which our experiences are derived?* We shall find that under three different forms there is always an irrational remainder, viz. in the relation of quality to quantity, in the significance which the time-relation has for the causal concept, and in the relation between the subject and object."[100] Although each of these three items is of interest, we will focus our attention on the third item, the relation between the subject and object. The reason that the subject/object relationship is of particular interest here is because of the central role that it plays in the development of the generalized complementarity framework we are using, in which the necessity to include a knowing subject is a central premise. We must explore the consequences of this premise.

In Kant's formulation, there is a fairly clean distinction between the subject, who supplies the forms, and the object, which supplies the content. Hoffding notes that this distinction is untenable, because the subject too is part of the world. The forms are not simply given but must be worked out during the process of creating knowledge. "In every cognition we can distinguish between *a subjective* and *an objective element*, between a knower and the thing known; both terms, however, are only given in mutual relation [...] we nowhere and at no time possess the pure Subject, with its forms, as an antithesis to a pure object [...] we really set up an objectively determined Subject (S_O) as the reverse of a subjectively determined Object (O_S)."[101] In other words, while our knowledge of the world is conditioned by our forms of perception and understanding, these very forms themselves are conditioned by the world we wish to have knowledge of. Absolute and complete knowledge can never be guaranteed under these conditions. Indeed, the problem is even worse than it initially seems, because the mutual subject/object relationship maintains itself recursively. "Each refers to the other indefinitely, and the irrational crops out in the fact that an infinite series (of the type: $S_1\{O_1\{S_2\{O_2\{\cdots)$ is both possible and necessary [...] the springs which feed the stream of thought are inexhaustible"[102]

These logical relationships that Hoffding explores can also take on a deeper existential meaning. The body is "from one side a thing among things and otherwise what sees them and touches them; we say, because it is evident, that it unites these two properties within itself, and its double belongingness to the order of the "object" and to the order of the "subject" reveals to us quite unexpected relations between the two orders. It cannot be by incomprehensible accident that the body has this double reference; it teaches us that each calls for the other. For if the body is a thing among things, it is so in a stronger and deeper sense than they [...] If it touches them and sees them, this is only because, being of their family, itself visible and tangible, it uses its own being as a means to participate in theirs, because each of the two things is an archetype for the other, because the body belongs to the order of the things as the world is universal flesh."[103]   Considered at this level, we see that our epistemological analysis is heading toward both an ontology that transcends human issues and at the same time introducing dimensions of meaning that are central to human concerns.

The Role of Objectivity

Before we go too far down this road, we should pause to think about the objections that a common-sense realist might have at this juncture. After all, how can it make any difference to the fundamental workings of the universe whether humans even exist or not, much less how they perceptually interact with it? Surely there must be rules (to be discovered by science) and some ontological ground that are independent of human consciousness. "Humanity is not the center or the measure of the universe or of reality. The universe is not dependent upon human perception [...] The quest here is for an objective picture of the cosmos as it would exist and function without the alleged contributions of human observers [...] In order to have knowledge and intelligibility regarding our transactions with the external world, some form of independent reality or objectivity is essential [...] physical reality is not partially created when the experiencing observer supplies the categories (as in Kant) [...] Reality is not constituted by, or contingent upon, human mental or 'observational' activities."[104] We meet here with a so-called "minimum metaphysics" that makes only those assumptions that are needed for clear thinking and seem obviously true. One of these important assumptions is the superiority of objectivity, taken to be so self-evidently correct that it requires no argument in its favor.

I am myself a strong supporter of objectivity. So is Hoffding. "The subjectivity of sense qualities, however, does not mean that they are invalid and unfit to guide us in the world. They stand constantly as tokens, signals, symbols, whose serial order we can point to as the expression of an objective series of events…"[105] The question is not whether objectivity is valuable, the question is whether objectivity is unlimited in its scope of validity and application. Hoffding's argument, as we've seen, is that our objective view of reality is constructed out of the elements that are bound by relationships of continuity and causality. He believes that we must always strive to enlarge this realm of objective reality and understanding, but that it will always face limits imposed by our own limited apprehension of Being. "…we run up against the irrational, and here perhaps we see most clearly how inexhaustible Being is in comparison with our knowledge […] Being may possess attributes that cannot be comprehended or defined by means of the dimensions in which our thoughts can move […] Knowledge, however rich and powerful it may be, is after all only a *part* of Being […] An exhaustive concept of reality is not given us to create."[106] The independent existence of the cosmos is not in dispute here. The dispute is concerned over what *we can know* about this independent cosmos. I find Hoffding's argument persuasive, and it follows from this that objectivity constitutes an invaluable but still-limited horizon; other horizons, grounded in apprehensions of other aspects of Being, may validly exist and we cannot dismiss the experiences of mystics, for example, as inherently inferior.

There *is* a manner in which objectivity is privileged, and the foregoing should not mislead us into accepting any assertion as equivalently valid as any other assertion. The formal rules of logic and rationality in combination with their empirical correlates still reign uncontested over their proper domain.[107] Incorrect or nonsensical propositions within this domain should be treated as ruthlessly as if the domain were subject to no limits or restrictions, and this includes propositions that claim some warrant based on revealed text, ecclesiastical authority, or a majority opinion. In applying the complementarity framework, we will examine carefully whether a proposition falls within this domain of objectivity or not—what I have previously referred as the conditions under which knowledge is acquired. The argument I have been developing claims that there are legitimate conditions under which knowledge is acquired such that this knowledge does lie outside this domain.

The Ontological Question

The believer in a mundane world still has two arguments left. The first argument is to claim that their horizon is still the only legitimate horizon, and that any other proposed horizon is illusory. But the only basis I can see for this claim is to deny Hoffding's outcroppings of irrationality and maintain that Being possesses only a continuity congruent with formal logic and rationality, and that the knowing subject has some transcendental status unconditioned by the world (i.e. the subject and object distinctly split). The relationship between the subject and object has already been dealt with in detail, and the assumption of the continuity of Being is difficult to maintain when even the continuity of formal logic itself has been shown to be incomplete. Godel's theorem has unambiguously demonstrated "that it is impossible to establish the internal logical consistency of a very large class of deductive systems—elementary arithmetic, for example—unless one adopts principles of reasoning so complex that their internal consistency is as open to doubt as the systems themselves."[108] If even the formal rules of logic are not self-contained and consistent, what grounds are there to assert that Being possesses these qualities?

The second argument, and perhaps the most difficult to refute, is to decouple epistemology from ontology and contend that what we know or don't know is unrelated to what is. Since the basis of my argument for complementarity has been primarily epistemological, this contention strikes a serious blow against it. On the other hand, what can we possibly *say* about the metaphysical ground of Being except what we *know* about it? Any positive statements that might be made, from the leanest minimum metaphysics to the most florid and grandiose systems, can be no more than dogmatism if they are divorced from our theory of knowledge. Hence, all of our arguments for complementarity on epistemological grounds must and do illuminate our ontological views. Statements like the following cannot be dismissed as invalid or irrelevant simply because they are not demonstrable within a mundane worldview: "Does doing away with the distinction of birth and death, for instance, in the liberated consciousness actually 'do away' with these 'realities' themselves? By realizing impermanence as the essence of everything whatsoever is one thereby freed from its bondage, not only psychologically but also ontologically? To answer this question leads us to the crux of the problem [...] Situating one's existence in the boundless dimension of being-nonbeing one realizes universal transitoriness as the only Reality, including himself in this realization. Reality is realized by him, who himself is a realizer of the Reality. This is an ontological, not psychological, awareness par excellance."[109] Of course this statement is amenable to critical scrutiny like any other, but I maintain that such scrutiny will only be meaningful within the complementarity framework advocated here.

The problem of Being and its relation to knowledge was also considered by Hoffding. He points out that since knowledge is always knowledge of relationships between things, then complete knowledge of a totality is inherently self-contra-dictory. If a totality is to be compared to something else, then it is by definition *not* a totality. Remarking that the problem is equally severe whether we restrict our concept of totality to empirical realms or broaden it to include the transcendental,

Hoffding notes that "the antinomy is the same in both cases. The irrational meets us here as it did in the problem of knowledge [...] In their different systems of thought, the philosophers have been too sure that Being in itself was a closed and constant totality...."[110]   After analyzing the problem in terms of various "type-phenomena" such as life, thought, matter, intelligibility, causality, plurality, and monism, Hoffding concludes that the unifying power found in Being will always be checked and limited by the irrational power found there too, and that the conflict between these opposing tendencies leads to the further development of a Being in the process of Becoming. He then explores the basis of the unifying tendency in Being, carefully considering both of the traditional sources: matter (materialistic metaphysics) and mind (metaphysical idealism).   Not only does Hoffding determine that neither of these solutions is adequate, he also notes that they do not necessarily exhaust all the possibilities. "...the difficulty would remain that matter could no more be derived from the psychical than the psychical from matter [...] but there is no proof that there is no other attribute in being besides these two [...] The empire of Being may be much vaster than the possibilities of our experience. Here, again, it is true that the world is great, but our mind is small; again we come upon the irrational [...] The possibility that there are more forms than our experience exhibits may signify that the whole problem lies deeper than has been supposed."[111]

Comments and Metaphors

It should be clear now that our ontology must be intimately connected with our theory of knowledge, and this implies that the complementarity framework we have developed based on epistemological considerations actually does apply to nature and not merely to our opinions thereof.   I would like to end with a metaphor, some final comments on the subject/object relationship, and a reconsideration of the analogical interpretation of Planck's constant **h** in our generalized framework. The metaphor I'm referring to is concerned with horizons. We've discussed the horizon of the person in a purely mundane world, and likewise discussed the horizon of the person in a purely sacred world. But these two horizons do not overlap or interact, and stepping from one horizon into the other would cause jarring cognitive dissonance. In arguing for complementarity of the mundane and sacred worlds, I implicitly reject both of these limited horizons. A good metaphor for complementarity might be a horizon in which one direction represents the mundane view and the opposite direction represents the sacred. Both views are thus within the same horizon, both accessible to the knower, and yet neither interferes with its contrary because they can't both be faced simultaneously. All you need to do, though, is turn around. A nice feature of this metaphor is that as you turn around, one view merges into the other and so there will be regions of overlap, corresponding to those difficult cases that require further analysis of the conditions of knowledge; these cases must certainly exist if we are to live in a single coherent world, and some of them will be given detailed examination in later chapters.

What is the subject/object relationship in each of these cases, the purely mundane and the purely sacred? In the case of the mundane view, as suggested previously, there needs to be a clean split between the subject and object. In its most extreme limiting form, this implies a complete separation and detachment, a sort of Cartesian mind that exists independently outside the world of things. As we've seen, Hoffding argues that such a mind does not and cannot exist, and Bohr argues further that the quantum postulate prevents such a complete separation even in the realm of physical science. As a limiting idea, though, it can have some usefulness in defining ideals of objectivity that may be well approximated in many situations. In the case of the sacred view, we have the opposite situation and the subject enters into a deep relationship with the world. The limiting case here is a complete merging of subject and object, occurring in various guises in the mystical literature of several religions as well as in some forms of existential phenomenology. Note that Hoffding's thinking would not allow this sort of merging to occur any more than a complete split. Once again, though, the limiting case is a useful ideal type to guide our thinking. As the knowing subject standing within the horizon defined by the complementarity framework turns toward different poles of the horizon, the knower's relationship with Being alters and in response Being reveals new and different aspects of itself.

Finally, let's take another look at the interpretation of the analogical generalization of Planck's constant $\mathbf{h}$ in the complementarity framework, in light of the new insights we've gained. Recall that the contingent physical fact that $\mathbf{h} \neq 0$ is in essence responsible for the limited subject/object relationship in Bohr's complementarity. The more profound subject/object relationship demanded by the theory of knowledge depends on no contingent fact; that relationship is built into our very role as conscious knowing agents. It is not avoidable. But, it might be imagined. What would the analog of $\mathbf{h} = 0$ correspond to in our imagination? In physics, $\mathbf{h} = 0$ corresponds to the classical limit, in which we have full and simultaneous knowledge of the exact values all dynamical variables, with no limitations on our use of space-time coordination and causality. We have complete knowledge of the system, and we don't need complementarity. The analog of this situation in our generalized framework, then, would be complete knowledge. The mundane and the sacred would be the same; for such an entity, for which I don't believe we can even begin to form a concept, scientific and spiritual knowledge would be identical. The direction such an entity gazed within its horizon would be all directions at once. It would be detached from the world and merged with the world, both at once. The entity would be omniscient. We don't have this option; hence, our need for complementarity.

Part III

Applications

10. Creation

Now that we have developed our ideas and methodology of complementarity as a way to understand nature in a broad sense, we wish to test whether it can provide any new insights by applying the general method to some specific problems and issues. We begin our series of applications in this chapter with an important and much-studied problem, the problem of origins.

How did things come to be in the beginning? This is a fundamental question for which almost every culture frames some sort of answer, typically in the form of a mythic story. These mythic tales serve more than one function, because in accounting for the origins of the world and its people they also provide both an account and a justification of the present order of things. Traditionally, cosmogony (account of the origins) and cosmology (account of the order of things) were linked together in a way that people understood. Our modern culture has lost these things, and in place of a creation myth that gives meaning to the present we have instead the science of physical cosmology. Physical cosmology does seem to be somehow related to our origins, yet it's clearly not an account of creation in any usual sense. Let's explore in more detail the real meaning of physical cosmology, the cultural role of creation myths, and the potential interrelationships that may exist between these differing attempts to organize our understanding of the world.

Physical Cosmology

Traditional usage of the word "cosmology" implied an underlying order that governed both the universe and humanity (along with any supernatural levels of existence, e.g. angels, that may occur), but the modern usage of "cosmology" in the sciences simply means that the object of investigation of that science is the entire universe taken as a whole. Physical cosmology is a rather recent addition to the sciences, dating from the early part of the 20th century when theorists tried to solve the equations of Einstein's general theory of relativity for the universe as a whole by making various simplifying assumptions. The choice of assumptions would specify some model for the universe, and the solutions of the equations would show how such a universe behaves and evolves over time. These early attempts were not informed by any data, but that problem was soon corrected. Present cosmological models are now tested by and based on a great deal of empirical information.

The theoretical structure upon which cosmology is based still rests on Einstein's 1916 general theory of relativity. This is essentially a theory of gravitation, but it's also a theory of space and time. Space, time, matter, and gravity are all indissolubly linked together; physics and geometry are no longer entirely distinct from each other, and space/time become physical entities rather than the inert container within which the laws of physics play themselves out. Space and time had already been melded into a single spacetime continuum by the special theory of relativity (in 1905), and now this spacetime became non-Euclidean, i.e. curved rather than flat. The curvature of spacetime, caused by the presence of matter, is what we have traditionally called gravity. The relationships among space, time, matter, and gravity are governed by a set of equations, and the application of these equations to the entire universe (along with their solutions) marks the beginning of modern cosmology.

Einstein performed such calculations, making a set of simple and reasonable assumptions about the properties of the universe. He assumed that the universe was homogeneous (the same everywhere, at least on a large scale) and isotropic (looks the same in every direction), consistent with the underlying ideas of relativity theory that there are no special preferred observational frames of reference. Any point in the universe is roughly equivalent to any other point, a statement that is often now called the "cosmological principle" and is still an important idea. Surprisingly, Einstein's solution of the equations under these conditions revealed that space was expanding with time; in other words, the solution was non-static. To obtain a static solution in which the universe remained the same over time, Einstein had to add an arbitrary constant to the equations. Work by de Sitter showed that this sort of non-static behavior occurs even in a universe without any matter in it. Extensive studies by Friedmann and by Lemaitre investigated a number of different models with varying assumptions, demonstrating that non-static solutions were obtained under a wide variety of circumstances. Although such solutions seemed perplexing at first, given the apparent static nature of the universe, observational astronomy soon revealed a surprising fact: the universe *is* expanding.

A number of astronomers, most prominently Hubble, were able to show during the 1920's that distant galaxies are receding away from us with a speed that is proportional to their distance. The velocities of such galaxies are relatively easy to measure using a phenomenon called the Doppler effect. The Doppler effect for sound is apparent in the lowering of pitch heard in a siren (e.g. an ambulance) that is traveling away from you (or a higher pitch if it's traveling toward you). These pitch changes are caused by changes in the sound frequency due to relative motion, and analogous changes occur in the frequency of light (in this case, the color of the light changes). Thus, the light from a galaxy that is traveling away from us has a lowering of its frequency and a shift toward the red end of the spectrum (hence the name for this effect, redshift). From measured changes in frequency (using spectroscopic analysis) the velocity of the galaxy can be calculated. Measuring the distance to such a galaxy is more difficult, but it can be accomplished by correlating brightness to some other physical property and then using the brightness to calculate the distance. Ultimately, a set of redshift and distance measurements were collected and the direct proportionality of velocity to distance was shown. Moreover, except for minor "local motions" it was found that every galaxy observed was in fact receding away from us. According to the cosmological principle, there is nothing special about our particular point in space, so we can conclude that every point is receding away from every other point. In other words, the universe is expanding.

Two important conclusions follow immediately from this fact. First, the expanding universe solutions found based on general relativity are dramatically vindicated. Second, if the universe is expanding as time moves forward, we can imagine moving backward in time with the universe getting smaller and smaller at earlier and earlier times. Clearly, there must be some limit to this process and as we extrapolate backward to earlier times the size of the universe would approach zero. The time at which this occurs might well be interpreted as the beginning of the universe (though there are many potential problems with such an interpretation). We will defer matters of interpretation until later, but the empirical fact of the universe's expansion clearly at least broaches the possibility of a finite age, an evolutionary history, and a beginning in time for the universe.

Leaving aside questions of origins, the cosmological

models based on general relativity taken in conjunction with the empirical data for an expanding universe strongly suggested that a very hot dense phase existed early in the history of the universe. The proposed existence of this hot dense phase offered the opportunity to add more physics to these cosmological models, because such hot dense conditions were expected to be just the conditions under which nuclei might be created (nucleosynthesis). This idea was explored by several people, with especially important early studies made by Alpher, Gamow, and Herman. Part of the problem is to determine properly the conditions found in the early universe (about 100 to 300 seconds old) and the other part of the problem is to calculate the nuclear fusion reactions accurately. It turns out that only the light elements were formed in the early universe, the rest being formed in stars (elements lighter than iron) and supernovae (the heaviest elements). At times that were too early, the temperature of the universe was too high for any nucleosynthesis to occur. As the universe cooled, a fair amount of helium was produced. But as the universe cooled further, there was not enough energy to create any heavier nuclei (except trace amounts) and the density also became too low for any further nucleosynthesis. The initial abundances of hydrogen and helium (plus traces of deuterium and lithium) are thus frozen into place and remain the same (except for subsequent stellar nucleosynthesis of heavier elements) to this day. Detailed calculations of these abundances have been made, quantifying the qualitative reasoning outlined here, and the results predict that about 25% of the universe (by mass) should be helium. This prediction is in excellent agreement with the measured data, offering strong support for this cosmological model.

Further dramatic support for the model came in the 1960's, when Penzias and Wilson accidentally discovered the cosmic microwave background radiation. As we've discussed, the universe cooled as it expanded. The very high-temperature radiation found in the early universe is still with us, but after billions of years of cooling and expansion its temperature is presently quite low, namely about 3 K (degrees Kelvin), i.e. 3 degrees above absolute zero. The energy of this radiation decreases as its temperature decreases, so it is presently at low energy and long wavelength, in the microwave region of the spectrum. As a consequence of our cosmological model, we would predict the presence of this radiation uniformly distributed everywhere in space, a fossil remnant of the early hot universe. Once again, we can calculate the properties of this radiation (technically, its blackbody distribution spectrum) very precisely and compare these calculations with very accurate measurements. The good quantitative agreement found plus the strongly intuitive explanation for the presence of the cosmic microwave background serve as extremely robust evidence for the correctness of the model. There is now virtually unanimous agreement that this model, known as the "hot big bang model," is substantially correct.

The Big Bang

It may be useful to keep separate the evidence, the coherent model that explains the evidence, and the narrative story that closely follows from the model. Often, only the narrative itself is given. Also, speculative embellishments to the basic narrative, such as inflationary scenarios, exotic dark matter, etc. are often included without distinguishing them from the well-established facts and consensus view. The familiar basic big bang narrative runs as follows: In the beginning, all of the matter and energy of the universe existed in a single point of infinite temperature and density. Space and time came into existence as this point started to expand; at the unimaginable temperature and density found in the early universe, matter and energy as we know them did not exist, being found instead in the form of an undifferentiated primal substance. The laws of physics as we know them likewise did not exist, all fundamental forces being merged into a single unified force. As the universe expanded and cooled, matter and radiation separated out while the unified forces sequentially broke into the forces as we now know them. As the universe continued its overall expansion, gravity accumulated some of the matter into stars and galaxies, which then evolved through their own life cycles, forming the heavy elements along the way. Elements up to iron were formed by fusion in regular stars, while the heaviest elements were formed in the explosions of supernovae. Eventually, these new elements found their way into later generations of new star systems, including our own, where conditions on the planet earth gave rise to the origins of life. Life then went on to evolve into the present global ecosystem, while the universe continues to expand, possibly forever (though no one yet knows).

The narrative is a little misleading in some respects, as is its name (big bang cosmology). It's called the big bang because it appears as though the beginning of the universe was a huge explosion at a specific point, from which the matter in the universe flew away with great speed and force. This seems to imply that the location of this initial point is the center of the universe, and that material is hurling outward through space. Both of these ideas are wrong. The cosmological principle assures us that there is no center of the universe; all points are equivalent. Moreover, space itself is expanding, and the matter is simply carried along with the expanding space. A huge explosion in space, hurling matter outward, can't possibly make sense because space itself was created in the initial event. This concept is extremely difficult to comprehend. "If we take any point in the present universe and trace back its history, it would start out at the explosion point, and in that sense the Big Bang happened everywhere in space."[112] Because we are trying to put inherently mathematical results into a pictorial language, we are generally forced to rely on analogies of some sort. "To use another common analogy, visualize the universe as the surface of a balloon on which the galaxies are distributed as spots. As the balloon is inflated, the spots will move away from one another. Clearly, there is no preferred position on the surface [...] Note that the center of the expansion (i.e., the center of the balloon) is not in the space-time geometry of the two-dimensional residents of the two-dimensional surface of the balloon."[113]

The big bang is the name for the initial event itself, and also for the cosmological model of which that event is a part. Ironically, the name "big bang" was coined by the opponents of the model in an attempt to discredit it. This initial event, however, despite the stunning successes of the cosmological model in accounting for physical observations, remains mysterious. As we extrapolate backward in time, we arrive at a point in time where the physics (namely, a proper theory of quantum gravity in which all forces are unified) is simply not understood. Although this time is *very* early (about 10^{-43} s after the initial event), we're not quite there yet. Even worse, the conditions at the initial event itself may defy any physical meaning at all: described as a singularity, it is a mathematical point with infinite density and infinite temperature. No laws of physics as we usually conceive them could exist under such conditions. What, then, is the meaning of the model now? Finally, there is the rather puzzling status of space and time, neither of which exist until the initial event takes place. If this is so, where does the singularity point reside, since no space exists for it to reside in? Likewise, our very language forces us to talk about conditions prior to the big bang, a contradiction in terms if time "began" with the big bang event. We are reminded of the comment that eternity does not mean "a very long time" but instead eternity means "the absence of time." These are mysteries, but they are mysteries we must confront if we are interested in Creation.

Creation Myths

Our craving for an understanding of our origins is deeply felt and virtually a cultural universal. Creation myths are not merely stories that satisfy a vague curiosity but rather are connected to fundamental issues of ontological import and existential being in the world. We are obviously not using the word "myth" in its frequent modern connotation as "an oversimplified primitive story" or worse yet "an outmoded idea now known to be useless and wrong." "Myth" in the present context means a shared narrative expressing truths not expressible by ordinary logical discourse. This is "myth as the medium for the articulation of our experience in the world and for the world's revelation of its own inner dynamic to the human mind [...] As long as we continue to be sentient beings we shall continue to need myth, since myth is the primary ground on which we articulate our experience of ourselves in our social and natural environment."[114] There are important myths concerning heroes, natural phenomena, cultural events, etc., but the myths concerned with origins, with creation, have a special power and significance. In part, this is because of the fundamental subject matter involved. Existence itself, the root of being, only exists by virtue of having come into being. How and why this happened can't help having profound importance. But even beyond this fundamental significance, the questions of origins are always tied with the questions of the presently existing order of things. The overall order of things, in turn, includes the role of humanity. "This parallel between the pattern of individual life-processes and that of creation as a whole also means that the character assigned to the cosmos determines the (inner) meaning it will have for man, and consequently the part he will play in it."[115]

Let's consider a few of the many hundreds of world creation myths. A very ancient example is the Babylonian account given in the *Enuma elish*. The primal father and mother (Apsu and Tiamat) exist as personifications of the original chaotic waters. From their union, two further generations of gods arise. These new gods begin to impose order on the chaos, but this process of creation is opposed by Tiamat, who vows to destroy them and reassert the rule of chaos. Tiamat is challenged by Marduk, a powerful warrior god of the third generation, and Tiamat (who now has the form of a giant dragon-monster) engages Marduk in single combat. Marduk slays Tiamat, dismembers her body, and uses its parts to create the sky and the earth. Marduk then continues to complete the job of creation, including creating humans from the blood of Tiamat's follower, Kingu. Scholars of myth have noted the possible psycho-sexual implications of this account, and also its interpretation as the victory of a patriarchal conqueror over a more ancient mother-goddess culture. A very similar set of themes are found in the Greek creation myth recounted in Hesiod's *Theogony*, where the familiar story of Uranos (sky father) and Gaia (earth mother) giving birth to the Titans is told. The Titan Kronos revolts against his father Uranos, killing him and usurping his power. But the son of Kronos, Zeus, in turn revolts against him and the gods led by Zeus eventually defeat the Titans, resulting in the reign of order (relatively speaking) experienced by early Greek culture.

The creation of the world from the dead body of a being, found in the *Enuma elish*, is a theme found in many myths. This theme is often referred to as the Ymir motif, named after the frost giant Ymir from the Norse creation story. The Norse gods, led by Odin, defeat the frost giants and create the world from his dismembered body. "In stipulating the use of a corpse to engender something new, the Ymir motif introduces the cycle of birth, death, and regeneration that is central to earthly life."[116]

The Ymir motif is found in a variety of cultures, including the Aztec (the monster Tlaltecuhtli), the Hindu (the giant Purusha), the Kabyles (the giant Ferraun), and several others. A variant is found in the idea of a being who commits a voluntary act of self-sacrifice, dying so that his body can be used to further creation. Examples of this are the Chinese myth of the giant Pan Ku and the Japanese myth of the creator deity Izana-gi.

A very different motif, but also one that is inspired by first-hand observations of the world, is that of the primeval egg. Within this egg lies all the potentiality of the cosmos. In one of the Hindu versions of creation, for example, a cosmic egg floating on the primal waters contains Brahma, who emerges from it after a year of meditation and then creates the sky and the earth from the two halves of the shell. A similar tale is told by the Samoans, whose deity Tangaloa-Langi broke out of the egg and used the pieces of shell to create the Samoan Islands. Sometimes a god resides inside the egg, while in other cases the world itself is born directly from the egg, as in one of the Orphic creation traditions. The cosmic egg motif is fairly widespread, including myths from the Egyptians, Tahiti, Tibet, Sumatra, the Finnish *Kalevala* epic, the Persians, and the Phoenicians. In some versions, a mythical divine bird or other creature lays the egg, and sometimes the egg is broken open by a violent action or event. In a variant of the usual theme, humans are sometimes the creatures that emerge from the egg.

A related theme is that of the primal procreative act. An archetypal Father God and Mother Goddess join sexually and give birth to the world. We have already seen examples of this in the myths involving conflict between successive generations of gods. Many other cultures also offer examples of the motif. In Egyptian mythology, for instance, the sky goddess Nuit and the earth god Geb are the offspring of Shu and Tefnut and are the parents of Osiris and Isis (also of Set and Nephthys). In the Maori creation story, the sky father Rangi and the earth mother

Papa give birth to six sons, who must separate their tightly joined parents and complete the final steps of creating the world. Here again, the psycho-sexual implications of such myths have been much commented upon, but they can also be interpreted more broadly in terms of the formation of dualities or polarities from initial unities and the rejoining of these fundamental polarities so as to create novelty and multiplicity. Such themes are found in Taoism and Cabala, for example, and the mythic renderings offer a more colorful pathway to the same ideas as the more austere religious philosophies offer. A variant of the primal parent motif is the case of human creation instead of world creation. The original parents of humanity may be divine, as in the Zuni creation story, or may be first-created humans themselves, as in the Navaho, Persian, and Biblical accounts.

Many of these stories don't seem to start "at the beginning" but instead seem to presuppose some already-existing starting point. Often this starting point is some form of chaos, and this undifferentiated state is sometimes identified with a vast expanse of uninterrupted water. The imposition of order on chaos as the basic scheme of creation was used by Plato in his literary version of creation, the *Timaeus*. But in some creation myths, the initial state is a void and existence must be brought into being by a deity using whatever method the story specifies. "In the myths from many cultures, creator deities think, dream, speak, or sing the cosmos into being. [...] for example, in the myths of the Laguna people of New Mexico [...] Thinking Woman conceives within her mind the original being of all that exists. Thinking Woman makes the world, including the thoughts and the names of all it contains..."[117] In some Gnostic creation stories, there is a sort of self-willing internal to God through which the world comes into being. Both in the creation stories of the Maori and in the Samkhya school in India, space forms spontaneously out of an initial void. The well-known account given in Genesis is usually interpreted as creation *ex nihilo,* but God does seem to bring the world into existence through the word ("Let there be light" and there was light). In the Cabala, there is a distinction between non-being (*Ain*) and the initial monadic point from which the rest of existence emanates (*Kether*), but any process by which the *Ain* gives rise to *Kether* is not specified. In the creation myths of the Zuni, "Awonawilona, the All-Container, is self-conceived. By His own volition, He comes into being, then *thinks* the outward forms of the cosmos, which until now is only void and black desolation. He becomes the sun, mists, clouds, and terra firma of the visible and tangible world."[118]

All of these myths are considerably more elaborate than the brief summaries I've given, and to really do justice to them they should be studied in more detail. The creation myths of complex and long-lasting civilizations like those of Egypt and India tend to be extremely elaborate, and the creation myths of virtually every culture contain far more than an account of origins. Explanations of various natural phenomena, both trivial (why the rabbit has short forepaws) and profound (why death exists) are often embedded within the creation narratives, along with explanations of the social order, moral order, and even eschatological expectations. The Persian creation myth of the Zoroastrians, for example, is set within the context of the divine battle between light and darkness represented by Ahura Mazda and Angra Mainyu. This ongoing struggle between good and evil dominates everything else found in the story. Also, the ancient cultural memories and unconscious archetypal projections that seem to lurk within the mythic stories can only be discussed thoroughly by examining the details of the stories themselves, despite the fairly obvious references to such themes that we've noted already in a broad general fashion. While we are not able to engage in such an in-depth study of creation myths here, we have probably seen enough at this point to carry out our more limited aim of exploring the relationship between traditional creation myths and modern physical cosmology.

Creation in Christian Theology

The most well-known account of creation in Western culture, of course, is the Biblical version found primarily in Genesis but also fragmentarily throughout other sections (Psalms, Isaiah, Proverbs, Job). The Genesis narrative still plays a powerful role in the thinking and beliefs of many people. Fundamentalists have made a literal understanding of this creation myth the cornerstone of their quarrel with modernity, rejecting both physical cosmology and biological evolution as false doctrines contradicting their beliefs. More sophisticated Christians who have embraced modern science while maintaining a strong commitment to their faith have adopted a variety of strategies, such as interpreting the Genesis creation account metaphorically, in order to reconcile and integrate their beliefs. Religious thinkers have also been quick to point out the limitations of the claims that science can legitimately make from a philosophical viewpoint, and sometimes note where these bounds have been overstepped by zealous promoters of atheism. An entire literature has developed consisting of theologians and scientists commenting on these sorts of questions, relating creation as given in Genesis to the results of modern science.

The significance of creation for theologians extends well beyond the explicit content of Genesis. The text is somewhat ambiguous about the question of whether existence came from nothing (*ex nihilo*) or was the product of order imposed on a pre-existing formless state. Christian theologians have decided this issue decisively in favor of *creatio ex nihilo* for a variety of reasons. In part, this doctrine was a reaction against the Greek and Gnostic competitors of early Christianity, emphasizing the goodness of creation and the distinct difference between God and the created world. Perhaps more importantly, the details of creation were less important than the message of salvation and redemption offered by Christianity, and creation *ex nihilo* is thought to be more consistent with these central doctrines. Creation *ex nihilo* "does not come initially from speculation regarding the origin of the cosmos. What provokes the idea is, in fact, the experience of divine redemption. It is the intra-cosmic experience of God's redeeming activity which leads eventually to the idea of God's act of cosmic creation."[119] The lesser importance of creation compared to redemption is consistent with the overall message of the Old Testament, in which the primary narrative concerns the covenant of God with Israel, as well as the New Testament. "It appears that God was worshipped as the redeemer of Israel before being worshipped as the creator of the world."[120] In addition, the conception of *ex nihilo* creation is tied to the power of God, who is not limited by the need for matter or by any properties it may have. The earliest attribution of much power to God (Yahweh) may have been related to the struggles between the Jews and their neighbors, who worshipped a variety of competing deities. Later Christian theologians made the limitlessness of the power of God a central tenet of orthodox faith, thereby lending more importance to *creatio ex nihilo* as a matter of doctrine. "Augustine argued that Divine creation is a far more radical relationship than mere making from materials already there. It is a total bringing to be, an act whereby the very existence of the world and of each thing in the world is affirmed and sustained."[121]

The response of Christian theologians to the results of physical cosmology has been varied. No less a personage than Pope Pius XII welcomed big bang cosmology as a vindication of *creatio ex nihilo* in the sense that a defined beginning in time of the universe is part of the theory. Although many qualifications must be added, some theologians still accept this basic reasoning. For example: "...for theologians to raise again the prospects of *creatio ex nihilo* understood in terms of a beginning to time and space is to be consonant with discussions already taking place within scientific cosmology [...] It simply makes sense these days to speak of $t = 0$, to conceive of a point at which the entire cosmos makes its appearance along with the spacetime continuum within which it is observed and understood. If we identify the concept of creation out of nothing with the point of temporal beginning or perhaps even the source of the singularity, we have sufficient consonance with which to proceed further in the discussion."[122]   Other theologians, however, including some who are also cosmologists, prefer to be more cautious in drawing any such conclusions. In some cases, they emphasize that the temporal beginning is only a minor point (in terms of religious significance) and not entirely well defined (if we recall Augustine's reasoning); the crucial point for such thinkers is the total contingent dependence of the world on God, true whether there is a beginning in time or not. Another fear expressed by theologians is that God will be employed only to explain what is missing in the theory, e.g. the ultimate reason for the singularity itself to exist; this is known historically as a "God-of-the-gaps" argument, and often backfires when further scientific progress fills in the gaps. Also, great care must be taken to maintain the proper boundaries between scientific and theological conceptions, issues, and languages in order to avoid confusion and/or drawing erroneous conclusions. "...it is obvious that certain key words have different meanings in cosmology, philosophy, and theology. Words like 'universe,' 'time,' 'space,' 'cause,' are cases in point [...] Rarely do people discussing a particular issue within an interdisciplinary context adequately make these crucial distinctions. Failure to do so, at the very least, prevents proper precision from being achieved and often leads to real confusion."[123]   On the other hand, careful and critical analysis can reveal interesting theological insights based on physical cosmology. For example, the grandeur revealed by cosmological science, the immense scales of space and time involved, and the rich variety of phenomena found all may serve as the basis for religious reflections on the greatness of God. Modern cosmology is also suggestive of a commonality and interdependence among all things. "We are part of an ongoing community of being; we are kin to all creatures, past and present [...] The chemical elements in your hand and in your brain were forged in the furnaces of stars. The cosmos is all of a piece."[124] These sorts of insights tend to be part of what might be termed a general theology of nature, however, and not specifically tied to any Christian doctrine, though surely entirely consistent with such a doctrine in many forms.

The Mythic Significance of Physical Cosmology?

The fundamental role of science is to bring order to our empirical observations, and we have seen how physical cosmology accomplishes this role. The cultural role of myth is more subtle and complex, involving an attempt to answer questions and bring meaning on a variety of different levels through a shared narrative. These two roles, though they are distinct, are not entirely divorced from each other. Science, if it is good science, almost always entails some sort of narrative structure and is used culturally as a source of meaning (the validity of which eventually requires rigorous and critical philosophical examination). Myth, on the other hand, has always functioned in part as a source of explanations for observed natural phenomena while at the same time being based in part on direct observations of the world (recall the primal parents and the cosmic egg, for example). Our task, then, is to understand more deeply how mythic and scientific pictures relate to each other while not confusing categories by blurring the distinctions between them.

In a sense, more is demanded of myth than of science. A creation myth needs to provide an overall meaning and context for our lives, addressing our values and our hopes as much as our understanding. Physical cosmology, of course, does none of these things. And yet, physical cosmology is globally accepted by all who subscribe to a mundane worldview; every creation myth is restricted to a local validity within some particular cultural and/or religious group. Might it not be possible for the narrative associated with physical cosmology, suitably interpreted and augmented, to supply the dimensions of meaning and value offered by a mythic account? This question deserves further exploration, and we will try to develop more insight by exploring it with the generalized complementarity framework that we've established.

Within this framework, any knowledge that is bounded by empirical and rational limits cannot contradict the mundane worldview, even if it is not part of that worldview. The narrative of physical cosmology, then, will serve to constrain the content of any sacred mythic narrative that can be found acceptable. This sort of constraint, however, is not really new and unexpected if we look at the content of traditional creation myths. Such myths are always constrained by observations of the natural world. Whether the first humans were made from cornmeal or red clay depended on the materials known to the ancient mythmakers of that culture; if we now know that humans are made from heavy elements forged by stellar nucleosynthesis, this is merely a more sophisticated version of the same underlying process. What we now lack is the mythic dimension accompanying the knowledge, but this lack bears no necessary relationship to the higher degree of refinement, correlation, and inference in the observations.

We can imagine trying to supply this mythic dimension by going in either of two opposite directions. Starting with the mundane outlook, we can develop the big bang narrative in the usual fashion and then extend the meanings possible within this narrative by invoking a sacred worldview that enlarges upon the restricted aspects initially inherent in it. Alternatively, we can start with a particular cultural creation myth with its own idiosyncrasies of narrative and associated meaning, and then proceed to reconcile this account with the mundane big bang narrative as appropriate; after repeating this process for a variety of traditions, a sort of global creation myth might emerge. Let's take a brief look at the possibilities found in both methods.

Starting in the mundane world, we've already seen how the astronomical data are made coherent and consistent with a big bang narrative. In the mundane world, this narrative can have no further meaning or implications. One question is this: what further and complementary meaning or implications might it have in a sacred world? Perhaps the first obvious answer to this question is the evolution of order from chaos. One of the overwhelmingly pervasive themes of many worldwide creation myths is the establishment of a reign of order from the initial chaos of the universe. Presupposing the validity and importance of this theme, we can obviously see its consistency with the standard big bang cosmology, in which the initially unformed primal substance gradually unfolds into matter, radiation, forces, space, and time as we know them. We can infer another layer of meaning from this by noting the implication of connectedness among all things based on their common origin in the single primal substance. Another direction for the apprehension of meaning involves the inference of purpose and design in the universe, but this is less directly related to either big bang cosmology or to creation as such, and we will treat that set of issues separately in a different chapter. Let's note in passing though the heavy element nucleosynthesis mentioned earlier, and the need for this process (or some such process, anyway) to occur in order for humans to appear in the universe.

The fact that it happened does not make it meaningful, but if we presuppose meaning and look for examples then examples are clearly not hard to find.

Working out a detailed account of creation as mythos grounded in a scientific big bang cosmology will not be further pursued here. Several such projects, with varying levels of detail and somewhat differing emphases, have already been conducted. Goodenough, for example, maintains a fairly naturalistic emphasis but adds a metaphorical overlay of sacredness to the purely scientific account.[125] A more imaginative treatment, with considerably more direct influence from traditional creation myths and religious ideas, has been given by Swimme.[126] Perhaps the most elaborate version of a creation myth (which actually unfolds into an entire religion) based directly upon big bang cosmology (in conjunction with other modern scientific results) is the work of Globus, which he expanded and developed over half a century and to which he gave the name "Veritism."[127] Efforts like this, which freely employ scientific results but in a radically unscientific fashion, are often ignored because they violate our usual categories and in so doing seem to be somehow wrong. Our task here is to look more closely at how these categories relate and to ascertain whether a work like Veritism, which is rather interesting and original in many ways, may have unsuspected philosophical validity based on a complementarity analysis.

Approaching the problem from the opposite direction, we may start with the premises found in traditional creation myths and try to arrive at a viable synthesis while taking into account the results of big bang cosmology. This approach, which I don't believe has ever been seriously attempted before, has several problematic features. For example, many of the myths from different cultures have mutually contradictory features that can't be resolved by appeal to the big bang picture; if these are left out to find a lowest common denominator, then we risk gaining nothing over the first method, but we have no criterion to make choices. For example, the myth of the primal parents has no place in standard big bang cosmology, but nor does this myth contradict anything in the cosmological narrative. A statement like "the Divine Being not only have a dual sexual nature but that They copulate on a higher cosmic level for the purpose of manifestation"[128] refers inherently to the *source* of the singularity that serves as the starting point of the big bang picture; the details of the picture itself can't tell us anything about this source. And yet, ascribing a fundamental polar duality to the Godhead (much less a sexual polar duality) clearly contradicts a number of other creation accounts. Likewise, creation *ex nihilo* is plainly inconsistent with the imposition of order on a pre-existing chaotic state, but both of these versions are equally consistent with the facts and their interpretation within big bang cosmology. The fundamental question to ask is whether this is a real problem or not, and the answer to this question depends on what we wish to know and understand.

Seen from the framework of complementarity, we have no compelling reason to reduce the rich worlds of mythos to a single self-consistent account. We are obliged to make the sacred and mundane worlds consistent with each other at the interface between them where they merge. Hence, physical cosmology might legitimately constrain some mythic claims (the chronological age of the earth, for example, to pick an ongoing contemporary bone of contention). But claims made within the realm of the sacred itself, without contact or consequence in the mundane world, have a wider range of validity and may well represent a set of complementary truths not necessarily bound by rules of logical consistency. This is not to say that we should abandon our quest for a coherent world-view or accept any claims uncritically, but the criteria for truth in the realm of the sacred need to be worked out on radically different terms. To say that the universe has existed eternally may appear to contradict the statement that it was created at a particular moment in time, but eternity (as well as time) is a deep mystery. An unambiguous claim of contradiction here seems to me to stem from an overly-mundane understanding of what these words and concepts mean. Ironically, an appeal to the mundane world, in the form of physical cosmology, gives us no unambiguous verdict, a result I find highly suggestive. A similar analysis applies to the other example discussed above, the issue of "primal parents" in the form of an underlying fundamental polar duality. In Taoism, to choose perhaps the clearest formulation, there is no inherent contradiction between the essential unity of all things in the Tao and the equally essential duality of all things embodied in Yin and Yang. The complex pantheon of Hindu deities, which also includes sexual polarities, does not conflict with the underlying oneness of all things. And in the Western traditions, the cabalistic approach includes a fundamental polar duality that stems from, returns to, and is ultimately indistinguishable from a monadic unity.

I believe that these sorts of insights, which can't be derived from cosmology under even the broadest kind of interpretation, are of crucial importance to a complete understanding of our lived experience in the world. Further, I would argue that complementarity warrants the need for these insights as well as the insights provided by the big bang picture. Just as importantly, our generalized complementarity formulation offers a way to mix the sacred and mundane pictures represented by myth and cosmology without sacrificing clarity of thought or mixing categories inappropriately, while at the same time preserving the richness of mythic thinking that is needed for it to serve its purpose in human thought, experience, and culture. Let's see how this might be done in a specific example, the Ymir motif.

I've chosen the Ymir motif because it seems to be mostly a primitive and bloody remnant of pre-rational minds, a particularly unlikely prospect for coherence with the modern scientific thinking of physical cosmology. The widespread appearance of this myth in many cultures and times suggests that it does, however, carry some important underlying meaning for humanity. What is this meaning, and can we map it onto other visions of reality as suggested by cosmology? If we look beyond the vivid pictorial imagery of slain corpses and particular organs or body parts morphing into geographical features, then we find archetypal themes of new forms arising from destruction, of life from death, of the sacrifice of some parts of creation in order that other parts may grow. Considering the facts recounted in the big bang creation narrative, these themes might well be seen in the death of stars by supernova to create the novel heavy elements in the universe or by the destruction of higher symmetries so that new forces can come into being. Of course, these are just scientific facts to which no meaning is necessarily associated, but by the same token no meaning can necessarily be prohibited. The ascription of meaning, the Ymir motif, to the death of a star in the creation of new elements is not merely a fanciful reverie, but nor is it an implication of the cosmological model. To interpret this meaning properly, we need to ask ourselves what the conditions were under which we came to this understanding. Presupposing the validity of the archetypal themes underlying the meaning of the Ymir motif (taken very broadly, not in the sense of details in specific stories), we come to understand this meaning within the context of a sacred world-view, where the universe is alive with spiritual potentiality. Subsequent to this understanding comes contact with the mundane world and a way of expressing the archetypal themes that's appropriate to the culture. In the primitive Norse culture, Ymir's blood came to form the life-giving water of rivers and oceans. We live in a radically different mundane world than the Norse tribesmen did, and the novel expression of these ancient themes in terms of our world's creation from the "corpse" of a dying star doesn't seem to be that inappropriate, even if our knowledge of this event was acquired under the rigorous conditions appropriate to scientific work. We must live in a mundane world to understand this process of nucleosynthesis, but we must live in a sacred world to understand that the process may also have deeper levels of meaning related to ancient and primal matters. Complementarity allows to acquire both levels of understanding and to keep them properly disentangled but *not* disengaged.

The foregoing analysis of the Ymir motif serves as an example of the kind of analysis that could be done for a wide variety of mythic themes associated with creation (the cosmic egg; primal parents; primal chaos turned to order; an uncreated initial void; good & evil, eschatology, and a moral order; redemption; the place of humanity in the created order; and so on). More detailed analysis of all these themes is beyond our present scope, but the methodology is fairly clear. Having made a point of contact between the mythic and the cosmological worldviews (associating the initial singularity with the cosmic egg, to note a rather obvious case), we then ask what it means to make that contact by looking closely at the entire worldview and what each term (singularity, cosmic egg) really signifies and under what range of conditions the signification is valid. If we are successful, we move from the stage of vague analogy to a stage of complementary understanding. To repeat a key point, the mythic dimension of understanding is only possible by accessing a sacred apprehension of the universe in general. It's worth noting that several of the mythic themes listed above (e.g. good & evil) will have little relationship to physical cosmology, which then serves mainly to limit the mythic picture (rather than generating novel imagery) but obviously can't even impose limits on aspects of the mythic narrative that are totally unrelated. Once again the complementarity analysis would be used to determine these relationships, areas of independence, and appropriate limitations.

Rather than continuing with such detailed analyses, I would like to end by looking more closely at two quite general issues. The first issue, minor but of longstanding interest, concerns the status of the cosmological principle. This cornerstone of modern cosmological theory implies that we occupy no preferred place in the universe, since it states that there *is no* preferred place. In the mundane world, it's clear that the cosmological principle is a sensible approximation and that, consequentially, we humans occupy no special position. It's often stated (or implied) that the cosmological principle has a metaphysical edge and that humanity has been dislodged from the central position it was thought to occupy during the years that Aquinas, Aristotle, and Ptolemy dominated European thought (a variation of the process is sometimes presented with a sequence of dislodgings starting with Copernicus and ending with Darwin). Something seems to be wrong with this assertion, however, if we consider that any individual human *does* perceive himself or herself to be the center of his or her universe. I intended that statement to be epistemological rather than cosmological, but ironically the cosmological principle actually lends some physical credence to our self-perception, because if the universe has no unique center then any point may be considered so, including the one I stand in. Actually, I don't believe that I occupy the center of the universe, but the argument I've outlined suggests that the metaphysical edge of the cosmological principle is rather dull. The proper way to look at this situation, I believe, is from a complementarity framework. In the mundane world, the cosmological principle is (as far as we know) correct. In a complementary sacred view, the role of humanity in the cosmos (using the word in the old fashioned sense now) is not determined by spatial position and this role may be very special indeed. Holding such a belief in no way contradicts the cosmological principle, and *vice versa*.

The second issue is more profound, and brings us back

full circle to the questions with which we started: creation, origins, the beginning. Why and how does the universe come into existence? This may or may not ultimately be a scientific question; we can't know what will be within the purview of a future science. One suspects, however, that the ultimate origin of being will always lie beyond scientific understanding. In any event, the present situation, in which the physics of the very earliest times is incomprehensible and the initial singularity an unrealizable physical state, certainly remains a mystery. And what does myth tell us to add to the discourse? We start with a Void, with a Word, with a Non-Being, with a Self-Created Being; in other words, we start with a mystery. This is the central ontological mystery, why there is existence instead of nothing. In a complementarity framework, you can try to penetrate this central mystery in the mundane standpoint of physical cosmology or the sacred standpoint of myth, in the singularity or in the eternal. We arrive here at the ultimate overlap between the sacred and mundane worlds, at the interface between them where they blend into each other indistinguishably despite the sharp differences in the directions we arrive from. The clarity of the scientific model gives way to puzzled contemplation of the initial singularity; the clarity of the symbolic meaning in the created order gives way to puzzled contemplation of the uncreated. Complementary visions of reality merge together at the beginning, at the origin, at Creation, because both visions of reality behold the same thing and it remains what it always is, a Mystery.

11. Mind and Brain

The existence of mind in a world of matter has been a deep mystery from antiquity until the present day. Idealism, Materialism, and Dualism have all made valiant attempts to solve this mystery and explain the relationship between the mental and the physical, between mind and brain as we would now say. No compelling explanation has so far been presented, and I certainly don't have one here. But, can complementarity offer us a method by which some new insights might be found?

In a mundane world, only brains exist and mind is merely a word to designate particular conditions of the brain. "... there is no such thing as the mind. There are certain activities of the brain [...] that it is convenient to consider as mental activities."[129] In contrast, a sacred world reveals that mind is intimately related to the soul. If these complementary views are somehow descriptions of the same world, then we might infer that soul and brain are related in some fashion that can be coherently explored. We'll eventually come back to this idea, but we first need to detour through the modern discoveries of neuroscience, cognitive science, and the philosophy of mind.

Anatomy of the brain on a large scale
On a large scale, anatomists divide the brain into a variety of areas, many of which have well documented functions.

The outermost layer of the brain is called the cortex, which takes up a greater fraction of the brain in humans than in any other species and is generally thought to be responsible for our greater self-awareness and linguistic power. The different parts of the cortex have many complicated relationships with each other and with other parts of the brain, so individual sub-areas are not solely responsible for specific functions; but having made this *caveat*, we can roughly attribute visual processing to the occipital lobes at the rear of the cortex; auditory processing to the area in front of this, on the sides of the cortex in the temporal lobes; language processing to the left temporoparietal area; motor and somatosensory functioning to areas around the top of the cortex; and association, learning, and attention to the frontal lobe.

Within the cortex are other parts of the brain such as the cerebellum, thalamus, hypothalamus, pons, and medulla. Whereas the cortex is more directly associated with cognition and conscious awareness, these inner parts of the brain perform vital functions at levels below our awareness. Some of our motor activity is directly willed, for example, but much of it is rather "automatic" like the feedback mechanisms that allow smooth movements, controlled by the cerebellum. Vital functions like breathing and heartbeat, not under our conscious control, are related to the medulla. Initial stages of sensory processing, before any information can enter our conscious awareness, occur in the thalamus. The hypothalamus is part of the limbic system, which is associated with emotions and learning. Other parts of the limbic system include the hippocampus, cingulate gyrus, and amygdala. In addition to all of these parts of the brain (plus many more that I've left out), which along with the spinal cord are collectively called the central nervous system, there is also the peripheral nervous system spread throughout our bodies carrying motor commands, sensory information, and control systems for various bodily functions. The entire nervous system is highly integrated, and all of the anatomically distinct regions listed above are interconnected to the others in complex ways. Any particular experience will involve most of these areas operating together in a systemic fashion.

Neurons, synapses, and brain physiology on a small scale

The small functional units that underlie the operations of the brain (and the rest of the nervous system) are single cells called neurons. Emanating from the cell body of a neuron are the filamentary branching strands known as dendrites and axons. The dendrites receive signaled information from other neurons; dendrites tend to be heavily branched and relatively close to the cell body. Axons send the signaled information out to other neurons, and although they vary in length (as well as in how much they branch) from one cell to another, some axons are extremely long. The axons from one cell terminate onto the dendrites of other cells at connection points known as synapses, but the manner in which information is transferred at the synapse is complicated. Leaving this complication aside for the moment, note that the pattern of interconnections for just one single neuron can be remarkably dense and convoluted. Since each neuron is connected into a network of other neurons with equally rich interconnections, the complexity of the pattern quickly becomes enormous even for a fairly small number of cells. Now note that the number of neurons in the human brain is roughly 10^{11}!

Another layer of complications involves how information is transmitted throughout the nervous system. The signals are transmitted along the axons by a mechanism of membrane depolarization. To understand this process, we must begin by realizing that in the neuron's normal state, with no signal propagating, the neuron maintains a potential difference (voltage) across its membrane. The inside of the neuron is about 60 mV negative compared to its environment; this is called the resting potential. The resting potential is partially maintained by the balance between concentration gradients and electrical attraction amongst various ions, including K^+, Cl^-, Na^+, Ca^{++}, and negatively ionized heavy proteins. Although this balance is maintained in part by differing membrane permeability for each ion, the Na^+ ions are also actively pumped outside the membrane by the cell. Thus the Na^+ ion concentration gradient is particularly high. The Na^+ ion channels through the membrane are gated in a way that is controlled by the voltage across the membrane. If the resting potential is disturbed by driving it more positive, and if this positive change exceeds a certain threshold value, then the membrane becomes temporarily permeable to Na^+ ions, which flood inside the cell driving it even more positive until there is even a reversal in the membrane polarity (i.e. the inside becomes briefly positive). The situation does not last long; within milliseconds, the membrane reestablishes its impermeability to Na^+ ions and instead becomes permeable to K^+ ions, which flood outside the cell to restore the resting potential. Meanwhile, however, the region of the axon near this event has now also experienced a slight positive voltage change (due to the event), and this change in turn drives another temporary positive spike in the voltage, which then induces yet another spike near itself in the same way, and so on down the length of the axon. This is called an action potential. An action potential, then, is basically a traveling positive voltage spike along the axon, caused by depolarization due to Na^+ ion transport across the axonal membrane. This is how neurons communicate with each other. But, what initiates the action potential in the first place? In other words, what is the source of the small positive voltage disturbance that causes the neuron to fire?

To answer this question, we need to realize that the gated

ion channels in the neuronal membrane can be controlled by chemical signals as well as by voltages. The chemical molecules used in this process are called neurotransmitters, and they are crucial to the functioning of the nervous system and brain. Recall that an axon terminates onto one of the dendrites of another neuron, forming a synapse. The synapse is actually a small gap (about 30 nm). When the action potential reaches the termination of the axon, Ca^{++} ions are induced to flow across the membrane, and these ions initiate a mechanism that causes a stored neurotransmitter to be released into the synaptic cleft. The neurotransmitter diffuses across the synapse to the dendrite membrane, where it binds to a receptor protein. The receptor protein, when activated by the neurotransmitter, then controls the operation of ion channels in the postsynaptic neuron, either directly (in a rapid process) or indirectly by altering other proteins ("second messengers") in a slower but longer-lasting process. In both cases, the result of changing the operation of the ion channels is a change in the membrane potential difference, contributing to the initiation of an action potential if the voltage change is positive. In this case, the action of this synapse is called excitatory, because it helps excite the firing of the postsynaptic neuron. Generally, several excitatory signals must be received simultaneously to reach the threshold for firing a neuron. But, the ion channels can also be instructed by the synaptic signal to make the inside of the cell even more negative, which opposes the kind of change needed to initiate an action potential. This will act to suppress the firing of the neuron, and so these are called inhibitory signals. Whether a neuron fires thus depends on the net effect of all the excitatory and inhibitory signals that it is receiving at a given time.

Whether the effect of any given synaptic event is excitatory or inhibitory depends on the neurotransmitter and receptor that are involved. Scores of substances have now been identified as neurotransmitters, and the effects they have can depend greatly on the kinds of receptors available at a particular synapse. A number of important neurotransmitters are amino acids, including glutamate, gamma-aminobutyric acid (GABA), and glycine. Glutamate is probably the most important and widespread excitatory neurotransmitter in the brain, and it has several kinds of receptors (the NMDA and AMPA receptors most prominently) that control different ions on a variety of time scales. GABA and glycine are inhibitory neurotransmitters, with GABA especially playing a major role throughout the brain. Some GABA receptors control chloride ions directly while others employ second messenger proteins. Another important class of neurotransmitters consists of amine molecules. This class includes acetylcholine, epinephrine, norepinephrine, dopamine, serotonin, and melatonin. Acetylcholine was the first neurotransmitter discovered, and it has essential functions in both the peripheral nervous system and the brain. Dopamine, serotonin, and epinephrine are monoamines, and these molecules are often referred to as neuromodulators because they work indirectly by affecting the operation of other neurotransmitters. "Unlike most other transmitters and modulators, the cells that produce monoamines are found in only a few areas, mostly in the brain stem, but the axons of these cells extend to widespread areas throughout the brain. In this way, a small number of highly localized neurons making monoamines can influence cells in many other locations."[130] A large number of psychoactive drugs appear to work by attaching to the receptors for dopamine and serotonin, either blocking or imitating their actions as the case may be. Lastly, a number of neurotransmitters are peptide molecules. These neuropeptides include the now-famous natural opioids, endorphins and enkephalins. A number of peptide hormones, such as oxytocin and vasopressin, also function as neurotransmitters. The distinction between a hormone and a neurotransmitter is sometimes blurry, but the hormones affect synaptic events by flooding the environment of the neuron (sometimes delivered through the bloodstream) rather than by being released from synaptic vesicles. The situation is made even more interesting when we consider that the hormonal action of the pituitary gland is under partial control of the hypothalamus in the brain. Obviously, the complexity of these neurochemical processes in the brain is astonishing. Several hundred molecules have now been identified as having probable roles in synaptic transmission, and we've seen a few small examples of the kind of tangled feedback mechanisms that are involved. The picture I've drawn is actually quite oversimplified, leaving out many interesting facts such as the purely electrical synapses that don't need chemical messengers, the many intricate steps involved in the second messenger processes, the myelin sheaths that increase axonal transmission rates, and so on. Rather than looking at such details, let's instead turn our attention to some examples of the specific brain functioning that arises from these neuronal processes and gives rise to perception, memory, and the other elements of our experience.

Mechanisms of brain functioning

Many aspects of the crucially important functions that our brains perform (perception, cognition, memory, emotion, motor control, and physiological regulation) are beginning to be understood. Visual perception, for example, has been studied extensively, and we now know a good deal about the processing pathways associated with vision. The lens of the eye focuses an (inverted) image onto the retina, but even at this stage some neural processing has occurred since the motion of the eye, dilation of the pupils, and shape of the lens all need to be controlled by neural feedback mechanisms. The image on the retina then needs to be converted into signals in the nervous system, a task accomplished by special sensory transducer cells known as rods (for low light intensity) and cones (for color perception at high intensity). Light-sensitive molecules (such as rhodopsin) in these cells undergo chemical changes that ultimately result, after a complicated chain of chemical reactions, in the initiation of a membrane hyperpolarization by blocking sodium channels. There is not a simple direct connection between the rods and cones and the brain. The rods and cones make multiple complex connections to so-called bipolar cells, which are interconnected to each other by horizontal cells. The output end of the bipolar cells connects to the ganglion cells, which might be thought of as the interface with the brain (the bipolar cell outputs also include lateral interconnections by means of so-called amacrine cells). The purpose of all this complexity is to perform a massive amount of information processing on the signals before they even leave the retina. Action potentials are first generated by the ganglion cells, and these action potentials are conveyed along the output axons of the ganglion cells into the brain. These afferent axons constitute what we usually call the optic nerve, and they terminate in an area of the thalamus known as the lateral geniculate nucleus. A smaller number of optic nerve fibers go to the superior colliculus and the hypothalamus, where the retinal information is used to inform motor functions and circadian rhythms below the level of visual awareness. The visual information itself is transmitted from the lateral geniculate nucleus into the visual processing area located in the occipital cortex. Though some go to other areas of the visual cortex, the majority of the axons from the lateral geniculate nucleus terminate in a striated part of the primary visual cortex, denoted area V1. Area V1 performs a number of basic visual processing tasks, discriminating shapes, position, orientation, and movement (possibly involving a kind of spatial Fourier analysis). Area V1 sends axons to other areas of the visual cortex, and these areas perform higher-order tasks such as the perception of forms, the motions of objects, and depth perception from stereoscopic cues. This is also the stage at which neural processing operates to produce our perception of colors, mostly in area V4. The cones themselves have a very broad and crude spectral response. This low resolution spectral information is heavily processed in the retina and in the lateral geniculate nucleus. The visual cortex further processes this information to generate the rich and varied set of hues and saturations that comprise our visual world. Finally, the information from the visual cortex is combined with other inputs (sensory, memory, emotional, etc.) to form our conscious experience, probably mostly in the frontal lobes.

Memory is another aspect of our mental functioning that crucially shapes who we are and how we think. The neural bases of memory are not fully understood by any means, but a great deal of progress has been made recently. Although memories are not stored in any particular localized region of the brain, certain brain structures are vital to the process of memory formation. "The hippocampus is most usually associated with learning and memory encoding (e.g. long-term storage and retrieval of newly learned information), particularly the anterior regions."[131] At the synaptic level, the brain is known to have a great deal of plasticity, i.e. synapses grow, form, wither, strengthen, and weaken as a result of experience. Dendrites and axons create new branches, and also lose existing branches. These kinds of processes are related to the formation of memories. For example, the successive simultaneous firing of multiple synapses on the same cell results in the strengthening of those synaptic connections (i.e. they are more easily fired in response to the next input signal), a phenomenon named Hebbian plasticity after the psychologist D. O. Hebb. A closely related phenomenon is long-term potentiation, in which a high frequency of firing at a synapse results in that synapse firing more easily in response to subsequent stimuli. The mechanisms of long-term potentiation have been studied extensively. In one case, the release of the neurotransmitter glutamate activates only the AMPA receptors under normal conditions, but a high rate of excitation can remove the Mg^{++} ion that usually blocks the NMDA receptor resulting in a much larger entry of Ca^{++} ions into the cell in response to the glutamate. These calcium ions initiate a cascade of protein reactions, including both phosphorylation and synthesis of several different protein molecules, by activation protein kinases. These proteins (a prominent example is CREB) then produce the long lasting changes in the behavior of the synapse associated with long-term potentiation. Good evidence exists to implicate mechanisms like this in the formation of certain kinds of learning and memory.

A number of other processes are coming to be better understood. One key aspect of conscious awareness, for example, is attention (what we are focusing on at any particular time and the mechanism by which the brain chooses what it attends to). Models for such attentional control have been proposed and compared to experimental findings, and a particularly important role seems to be played by the anterior cingulate gyrus, a part of the frontal lobes. The attentional control networks are distributed throughout the brain, with the anterior cingulate gyrus playing the role of coordinator in such networks rather than issuing commands. "Anterior cingulate connections to limbic, thalamic, and basal ganglia pathways would distribute its activity to the widely dispersed connections we have seen to be involved in cognitive computations [...] computations appear to pass information back and forth to coordinate their results."[132] This kind of computational model, characteristic of the discipline known as cognitive science, has also been applied extensively to the problem of language acquisition and use. Cognition itself is often reduced to deployment of language algorithms in this kind of approach, and a large body of work exists in the area. "The basic idea of cognitive science is that *intelligent beings are semantic engines*—in other words, automatic formal systems with interpretations under which they consistently make sense."[133]   This kind of information processing paradigm is especially powerful when it can be combined with real data about how the brain operates in terms of neurophysiology and neuroanatomy, and a good deal of progress along these lines has been made. Most of this progress has been made in the areas of cognition and perception, not in other areas such as emotion. Research on emotion has mostly been limited to primitive instinctual responses like fear, and within this restriction much has been learned. In particular, the role of the amygdala in the fear defense system has been studied extensively. The lateral nucleus of the amygdala receives sensory inputs directly from the thalamus and again from the higher cortical areas. The less-processed inputs from the thalamus offer a much faster response (to a loud noise, for example) but no detailed information about the threat (or lack thereof). The processed information from the cortex is slower but offers more accurate knowledge of whether the threat is real or not, all underneath the level of conscious awareness. These few examples of our understanding of brain functioning could be expanded considerably, but they will suffice to typify the sort of knowledge we have.

Neuroscientific tools

How do we know these things concerning the workings of the brain? More precisely, what do we know and what do we infer? These narrative descriptions of how the brain works, which we have been considering, are all basically coherent explanations of indirect data obtained from a variety of sources. Each source of data provides information for a limited range of time scales and spatial resolutions, and the data needs to be interpreted appropriately.

Basic anatomical information, both large-scale and

microscopic, provides a foundation to build on. Microscopic anatomical information is greatly augmented by the use of stains and dyes with optical microscopy, and extremely high resolutions are possible using electron microscopy. Tracing the complicated networks of neuronal connections is also accomplished by injecting dyes and radioactive tracers into neurons, which then transport these materials throughout the axonal and dendritic pathways. None of these methods can be performed *in vivo*, however, and hence provide no physiological information.

To get information about the actual functioning of the nervous system, a variety of methods are available. Electrical recording of neuronal activity has traditionally been an important tool. The overall electrical activity of the human brain can be measured by placing electrodes on skin outside the head, and this method has the ability to record activity on a rapid time scale (ms), but it lacks spatial resolution (because the cerebrospinal fluid is fairly conductive) so the data represents the average activity throughout the entire brain. The most useful measurements have been those that employ a lot of averaging and compare the electrical signal to some specific stimulus, producing a so-called "event-related potential" (ERP). Electrical recording can also be performed on single neurons by using microelectrodes, but only if the neurons are accessible (usually in animals or cultured tissue samples). An early example of this method was the exploration of the action potential using microelectrodes inserted through the membrane of the giant squid axon. More recently, the method was used extensively to study long term potentiation and to look at the functioning of other neuronal circuits. In the case of single cell recordings with microelectrodes, the spatial and time resolutions are both excellent, but the problem is that *only* microscopic single-event information is available; the functional operation of any single neuron is only meaningful taken in relation to the functioning of the other neurons making up the circuit of which it's a part. Also, the need to have physical access to the cell is clearly a disadvantage. Another non-invasive technique to study the human brain's electrical activity is to measure the magnetic fields that are produced. This method offers higher spatial resolution than studying ERPs, but it is very difficult to measure such tiny fields and to interpret the data after it's measured, so this technique has not been used a great deal. Also, electrical activity is not the only important data, since chemical neurotransmitters play such a key role in neural functioning.

To explore those aspects of synaptic functioning, drugs are administered that block or imitate the role of some specific agent, and the effects of this intervention are observed. For example, a chemical called amino-phosphonovaleric acid blocks the NMDA receptors, and the famous poison curare acts by attaching to acetylcholine receptors and blocking their action. The mechanisms can also be more complicated than attaching to a receptor; a chemical might enhance the release of a neurotransmitter or amplify its effect, or the reuptake of a neurotransmitter back into its vesicles after a synaptic firing might be blocked. All of these mechanisms can serve as tools to study the workings of the synaptic junction, with behavioral changes, electrical measurements, and chemical analysis all used to ascertain the effects of administering the pharmacological agents. Sometimes these agents may even destroy the neural tissues they affect, but the observation of effects due to destroyed areas of the nervous system (lesions) has also been a source of information when the lesions occur naturally. In humans, this destruction can be caused by accidents and diseases, for example, and the behavioral effects correlated with the anatomical areas affected. One famous instance of this method is the research on "split-brain" patients with a severed corpus callosum, but many studies have been done attempting to locate the parts of the brain that store memories or integrate the personality and so on. All such studies have severe limitations (the lesions can't be controlled, most brain functions are not really localized, and the tissue can't be examined until after death).

A recently developed tool for studying the human brain non-invasively is the use of modern imaging techniques such as the CT (computed tomography) scan, MRI (magnetic resonance imaging), and PET (positron emission tomography) scans. These methods have spatial resolutions down to about a mm and can collect data within a few seconds of a stimulus under favorable conditions (usually longer, though). Along with anatomical information, scanning techniques (such as PET) can measure metabolic rate information, telling us which areas of the brain are using more energy during the accomplishment of some cognitive or perceptual task. It's good to remember that the information is indirect, however, and may be telling us very little about the processing mechanisms of interest. Indeed, this point is true of all techniques, and in practice we must combine pieces of information from disparate sources in order to make a coherent narrative. Single cell electrical recordings from hippocampal *in vitro* tissue, studies of long term potentiation and NMDA receptors in the sea slug *Aplysia*, and the effects of hippocampal lesions on human memory are all combined to formulate a model for how memories are formed. The models are interesting, and we have good reasons to believe that neurochemical processes occurring in sea slugs and glass dishes really are similar to those occurring in our living brains. But a model should never be taken too literally, and what the model doesn't tell you should always be considered.

Mind, self, and consciousness

The various neuroscientific models we've looked at so far all explain some interesting things (and leave open some interesting questions) within the framework of what such models might possibly be able to explain. Now we will expand our horizon a bit and consider whether such models have intrinsic limitations on what they even possibly *could* explain, and what those limitations might be. To start to grasp what these models leave out, consider your own first-hand conscious experience of the world. All the interesting facts we learned about the visual information processing circuitry of the eye and brain, for example, don't ultimately translate into your direct visual experience of the seen world. Our knowledge concerning the role of the anterior cingulate gyrus in controlling attention doesn't offer you any new insights into how you are making the decision, right now, to pay attention to this text or to something else. What we learn about is the operations of electrochemical neural circuitry; what we experience is our conscious awareness. How does our conscious awareness arise from the working of our neurons? This is still a mystery, on several different levels, that's not effectively addressed by the methodology of cognitive neuroscience. The discipline that tries to take this mystery seriously is usually referred to as philosophy of mind.

We have now introduced a distinction between mind and brain. Mind is what experiences the world and has self-awareness. The brain is clearly related to mind in some important way, but the brain is not the same as the mind. This assertion, incidentally, is by no means accepted by everyone, and we will explore the issue more thoroughly as we proceed. For now, consider this line of reasoning: "...where do we situate the qualities of experience? Your first instinct was to locate them in the brain. But inspection of the brain reveals only familiar material qualities. An examination of the brain [...] reveals no looks, feels, heard sounds [...] The idea that these qualities reside in your brain, then, appears unpromising. But now, if the qualities of your experience are not found in your brain, where are they? The traditional answer, and the answer that we seem driven to accept, is that they are located in your mind. And this implies, quite straightforwardly, that your mind is somehow distinct from your brain."[134] There are a variety of strategies to counter argumentation of this sort, but most of them rely fundamentally on the presupposition that the qualities of experience are not genuine phenomena and therefore require neither an explanation nor a mind to reside in. We'll examine several variants of this position more carefully later.

There is another even more radical assertion that we might make by following this line of reasoning. If we accept that the mind is the seat of conscious experience and awareness, and if we accept the distinction between the mind and the brain, we may then ask where the mind is located. Further, we may ask what the mind is made of. Assuming that the mind is neither located in the brain nor made of the material the brain is made of, we are left with few other options to answer such questions. It becomes difficult to avoid the conclusion that the mind is not a part of the material world at all. Mind, in this view, is a non-material entity with properties that are different from any material thing. Such conclusions, concerning mind, have indeed been asserted by several philosophers over the centuries (and including the present day) in several different formulations. For example, "the failure of logical supervenience directly implies that materialism is false: there are features of the world over and above the physical features."[135] The "features of the world" referred to here are basically what we call mind, and a large amount of subtle argumentation is offered to substantiate this stated conclusion. Materialists have mounted some strong counterarguments, though in certain cases some of these materialist arguments either miss the point or else look suspiciously close to the views they are attacking.

The ontological issues require considerable nuance in their handling, though, and fine distinctions must be made.

"In contemporary philosophy of mind, substance dualism has largely been abandoned [...] Thus, *ontological physicalism*, the view that there are no concrete existents, or substances, in the spacetime world other than material particles and their aggregates, has been a dominant position on the mind-body problem [...] the only substantive remaining issue concerning the mind-body relation has centered on *properties*—that is, the question *how mental and physical properties are related to each other*. Here, the main focus of the debate has been, and continues to be, the controversy between reductionism and non-reductionism."[136] In other words, the mind (in this view) is not something apart from the material world but instead is a novel property not predictable from and unlike any other property of the material world. Hence we see that philosophers of mind holding exceedingly similar positions might be categorized as either materialists or non-materialists in their understanding of the place of mind in the world, the difference turning on whether the acceptance of a so-called "property dualism" for mental phenomena is tantamount to a rejection of materialism or not.

A related issue concerns the distinction between logical supervenience and nomological supervenience. Supervenience means, very roughly, dependence upon or determination by. The conductivity of a metal, for example, supervenes on the electron states of the individual metal atoms. But although my first example suggests reductionism, supervenience does not necessarily entail reductionism. The existence of a strange attractor in the motion of a forced, damped pendulum supervenes on the physical details of the pendulum design, but the strange attractor is not reducible to or derivable from these details. Or consider this example[137]: the beauty of a painting supervenes on the exact distribution of the paint on the canvas. But the beauty of the painting is not reducible to the paints, it is something extra. In the present application, we may say that mind supervenes on the brain (plus the rest of the nervous system and body). In this way, we retain some sense of a materialist ontology while admitting that the mind is not just another routine part of the material world but instead entails novel properties. The foregoing is nomological (lawful) supervenience because it tacitly assumes that the laws of nature are true. Logical supervenience would make the stronger claim that mind supervenes on the brain in any conceivable universe; identical conscious awareness *must* emerge from an identical body/brain in the way that a circle *must* contain points equidistant from a center not because we happen to live in a world where that's true but because it's a logical contradiction for it to be false. Those who attach great importance to logical supervenience (and agree that the mind does not logically supervene on the brain) will consider the case against mental events being a part of the material world to be a stronger case.

As interesting as these technical points in the philosophy of mind may be, most of them are pressing issues primarily due to the presupposition of an otherwise materialist ontology to start with. Although traditional Cartesian substance dualism appears to be an almost universally rejected idea, there might also be novel (or even old traditional) alternatives both to Cartesian dualism and to a monism grounded entirely in scientific materialism. We'll look at some of these alternatives later, and consider whether they might be part of a viable and coherent metaphysics within the complementarity framework. The present considerations based on contemporary philosophy of mind, like the previous neuroscientific findings, will provide valuable input into that examination. Before we engage in this further widening of our horizons, however, let's take a closer look at some of the particular formulations devised within philosophy of mind.

Duality, identity, functionalism, and so on

Since many of the ideas we'll look at are presented using the Cartesian mind/body split as a kind of foil, let's briefly consider Descartes's concept. In this view, the mental and the material are two fundamentally different kinds of substances. Material substances have spatial extension, in contrast to mental substances, which do not. Material objects are located in space, but mind has no particular location or other spatial attributes; no material object, on the other hand, is capable of thinking. This conception leads to the infamous mind/body problem, because our bodies are material objects and they don't seem to be independent of our minds. A number of corollary problems follow: How does our mind, made of a different and incommensurable substance, affect (or be affected by) our body. How can mental events, operating outside the material world, have causal consequences within the material world if this world is considered to be causally closed and operating according to lawful patterns of interaction? These kinds of problems have resulted in Descartes's concept, usually called substance dualism, having few proponents for the last couple of centuries. While we can get rid of dualism by postulating that everything in the world is fundamentally mental (a view known as Idealism, which has had several champions over the years), we've already noted that the more popular form of non-dualistic metaphysics these days is scientific materialism. Let's look at some attempts to locate the mind in such a system.

The most influential school of thought around the middle of the twentieth century was the so-called identity theory. Identity theory claims that our consciously aware experience at any given point in time is nothing more than the specific detailed neurophysiological state of the brain at that point in time. In other words, our mind *is* our brain; nothing more and nothing less. Identity theory claims, justifiably, the virtues of parsimony and simplicity. In addition, identity theory certainly does get rid of dualism. The argument for identity theory is often presented in the form of an analogy: we can speak of lightning strikes or speak of electrical discharges, but we are not speaking about two different things, only one single thing because lightning is identical to an electrical discharge. In the same way, we might speak of our conscious experience as well as of our brain states, but we are really only speaking about one thing. So why include both? There is no particular mystery to solve in this view. "Once we rid ourselves of the phenomenological fallacy we realize that the problem of explaining introspective observations in terms of brain processes is far from insuperable."[138]   The argument may be interpreted in two ways, though, and one of these ways claims much more than the other. That any particular mental state (experienced awareness) bears a one-to-one correspondence to some specific neurophysiological configuration is plausible enough in principle (albeit not necessarily of any particular practical use). But the stronger claim that this truism essentially implies the nonexistence of experienced awareness as a category separate from the neurophysiological configuration is certainly not compelling. Our experienced awareness is precisely the primary data that we wish to explain by reference to the brain functioning, and saying that this awareness simply *is* the brain functioning explains nothing at all. Assuming that we wish to take mind and conscious awareness seriously, identity theory seems to beg the question of how these things arise. Somewhat surprisingly (at least to me), however, the major objections to identity theory are not based on this type of reasoning. Instead, the main objections to identity theory are based on the idea of multiple realizability.

The central premise of the multiple realizability argument is that the neurons themselves are not the important aspect of brain functioning, but rather it is the *organization* of the neurons that plays the vital role. This view is generally known as functionalism. Once again, the basic idea here is rather plausible. After all, the neurons of most species operate in very similar ways; what separates a human from an earthworm isn't having a radically different kind of neurons, it's having a radically greater complexity in their number and organizational interconnections. But the claims of functionalism go well beyond such common sense considerations. The next step of the argument is to point out that this organizational pattern doesn't necessarily need to be realized in a human nervous system. Since it is the organizational pattern as such that plays the key role, this pattern might be realized in any number of different material substrates, and in every case the same pattern would give rise to the same conscious awareness regardless of the material realization (hence the term "multiply realizable"). The analogy that is often made uses a computation performed by some sort of device. The device may be made of many different materials: transistors, vacuum tubes, gears & levers, beads on wire, or perhaps neurons. This material manifestation is unimportant, because the truly crucial thing going on here is the algorithmic operations needed to perform the computation and these operations are the same in every case. The differences are in how these operations are realized by a concrete material system, but these inessential differences don't matter for the result since the essence of the computation consists of the operations themselves. We might say that the software can be multiply realized in a variety of hardware choices, but it's always the same software. Note that the organization and operations (software) have taken on a kind of ontological status independent of any particular realization (or, indeed, any material realization at all). The main point of functionalism is that we should associate mind with the (multiply realizable) software, which happens to be realized by the hardware of our brains.

An extensive discussion of functionalism soon leads to

the consideration of Turing machines, Searle's Chinese room argument, and other arcana dealing with the question of whether a machine can think. I won't undertake such a discussion here because the material is widely available elsewhere and it would take us too far afield from our main purpose. It's worth briefly noting that the question comes up so often because functionalism underlies much of the work done in the influential disciplines of artificial intelligence and cognitive science. Cognitive science, although it combines elements of psychology, linguistics, artificial intelligence, and neuroscience, is primarily influenced by the paradigm of thinking as computation. Functionalism provides an ideal philosophical underpinning to this approach. Arguably, both functionalism and cognitive science represent the dominant way of conceptualizing mind/brain issues today. The major drawback to this approach, I believe, is that it tends to limit our focus to those aspects of mind that are amenable to algorithmic analysis. Another important problem is that although realizations of the organizational pattern underlying mind may indeed be independent of their material substrate in principle, in actual practice the details of the implementation have proven to be decisive, a fact which has been the bane of the strong AI movement. Only in science fiction has this problem been overcome. On the other hand, functionalism is an intriguing idea, and even an appealing idea if you have a certain predilection for the abstract. If you restrict your discourse to imaginary situations and construct them carefully enough, functionalism can overcome many of its drawbacks. I will have some possible uses for the functionalist point of view later.

All of these viewpoints share certain shortcomings. One issue is the limited contact they seem to make with the facts of neuroscience, which offer a rich array of empirical insights into the workings of the brain that ought to inform us (I would think) about the workings of the mind. At the same time, in contrast, a second issue is the impoverished conceptualization of mind and self that they seem to be trying to explain. The richness of our emotional lives, aesthetic insight, creativity, and so on don't seem to exist in a discourse that "reduces mental properties to causal powers."[139] In fact, the magnitude of this shortcoming is demonstrated by the role played in philosophy of mind by the so-called "problem of qualia." Qualia are the phenomenally experienced perceptual sensations: the blue of the sky, the red of the tomato, the smell of the perfume. The qualia are the actual experienced feels, sounds, sights, smells, and tastes, the raw material of consciousness. Yet, in most of these philosophical theories of mind, qualia don't exist. There is a raging debate over whether this means that the theories are deficient or that the qualia really must not exist because there is no place for them in the theories. I wish to suggest that the existence of this debate implies that we need to broaden our horizon yet again, and we are about to do so.

A radically different view of mind

"....not only does the capacity of our total consciousness far exceed that of our organs, the senses, the brain, but that even for our ordinary thought and consciousness these organs are only their habitual instruments and not their generators. Consciousness uses the brain which its upward strivings have produced, brain has not produced nor does it use the consciousness."[140] We see in this statement a very different approach to the questions we've been considering, starting with the assumption that the material world is not the sole aspect of reality; in fact, it's not even the primary aspect. Based on traditional Hindu metaphysics combined with the spiritual experiences of the author (Aurobindo), this view makes the primary aspect of reality an underlying spiritual unity termed "Sachchidananda." Similarly to the Trinity in Christian metaphysics, Sachchidananda is One and also Three, namely existence, consciousness, and bliss. Consciousness in this usage obviously has a different meaning than we've been assigning it so far (individual experienced awareness). Here, consciousness is a universal property that imbues all things, material and immaterial alike. The consciousness of an individual human being in this view is a limited participation in this universal consciousness. All beings and things, animate and inanimate, partake in this universal consciousness to the extent possible given their mode of existence in the material world. While I wouldn't presume to try to describe or explain this consciousness aspect of Sachchidananda, I am confident that I can say that it is nothing like our ordinary human consciousness and undoubtedly incomprehensible in terms of any rational categories we might construct. The key point here is that our own minds don't emerge from the processes of unconscious matter but instead are material manifestations of an already existing mentality in the world.

This idea of the cosmos having an underlying substrate of mentality is not unique to Hindu philosophy by any means. "Greek thinkers regarded the presence of mind in nature as the source of that regularity or orderliness in the natural world whose presence made a science of nature possible [...] They conceived mind, in all its manifestations, whether in human affairs or elsewhere, as a ruler, a dominating or regulating element, imposing order first upon itself and then upon everything belonging to it [...] The life and intelligence of creatures inhabiting the earth's surface [...] represent a specialized local organization of this all-pervading vitality and rationality, so that a plant or animal, according to their ideas, participates in its own degree [...] intellectually in the activity of the world's 'mind', no less than it participates materially in the physical organization of the world's 'body'."[141] Here again, mind is conceived as a primary entity that organizes matter and eventually manifests itself as the mentality of organisms that are made of matter, in contrast to the presupposition of the previous sections that matter is primary and that mind supervenes on it. Another variation of this theme, stressing a holistic underlying unity, is found in Chinese thought. "Indeed, the dichotomy of spirit and matter is not at all applicable to this psychophysical structure. The most basic stuff [*ch'i*] that makes the cosmos is neither solely spiritual nor material but both. It is a vital force [...] We want to know in what sense the least intelligent being, such as a rock, and the highest manifestation of spirituality, such as heaven, both consist of *ch'i* [...] The uniqueness of being human, however, is not simply that we are made of the same psychophysiological stuff that rocks, trees, and animals are also made of. It is our consciousness of being human that enables and impels us to probe the transcendental anchorage of our nature [...] The internal resonance of the vital forces is such that the mind, as the most refined and subtle *ch'i* of the human body, is constantly in sympathetic accord with the myriad things in nature."[142] In one form or another, ideas similar to this (in which the source of mind exists outside a mundane materialist framework) can be found in many times and cultures, including examples such as the writings of Plotinus, Avicenna, Ficino, and Whitehead.

There is also a different, but not unrelated, strand of thought that grounds the human mind in the existence of a divine Mind. Here the greater Mind is outside the material world rather than inherent within it, but the element of something in mentality that lies beyond the material is similar in both views. Thus, we can refer to "the existence of natural beings who at the same time transcend the natural level because they are rational"[143] and consider this rationality of the human mind to be a reflection of and partaking in the divine Mind. To examine the details of these many different conceptualizations is beyond our scope, but the existence of such a rich literature, at least some of which is contemporary, invites us to ponder whether the presuppositions of neuroscience and conventional philosophy of mind may be too limiting. We'll return to this question in more detail and consider the degree to which it's possible to take these ideas seriously in light of modern knowledge, and what conclusions might be reached if we do, later. First, we must consider another ramification of our broadened outlook concerning the mind, namely the relationship of mind and soul.

What might soul mean?

Words like soul, mind, spirit, and psyche are not very sharply defined. Not only do these words have overlapping connotations that merge into each other, but they also change over time, vary with culture, and present formidable problems for translators of texts from these times and cultures. In the ancient Hebrew of the Old Testament, *nepesh* might mean vital force, breath, desire, or even self. This word was subsequently translated and retranslated serially into *psyche* (Greek), *anima* (Latin), and soul (English). The Hebrew word *ruach* (usually translated as *pneuma* and *spiritus*) also broadly meant life-force or breath, but its connotation tended to be more general and less associated with an individual human. What we would call mind, however, is most closely associated with the Hebrew word *leb*, which literally translates into English as "heart" since this organ was considered the source of memory, emotion, cognition, and intellect. All three of these attributes (*nepesh, ruach, leb*) were integral parts of what it meant, beyond the existence of flesh, to be a human being.

Similar kinds of overlapping usages are found in the ancient Greek texts of Homer, where *psyche* (life-force), *thymos* (source of emotions), *nous* (mind), and *menos* (soul as substance) are all employed in a variety of different contexts with related shades of meaning. "Of the three terms used by Homer in descriptions of individual souls, the term most closely associated with what we now think of as higher-order cognitive functions was *nous* [...] which we might perhaps render as 'intellect'."[144]   There was no single term that precisely and uniquely meant either soul or mind, but instead there were many terms that contained elements of both. There was one feature of all these conceptualizations of the soul, both Hebrew and Greek, that they held in common, namely that all of them tied the soul to the body in some way. An immortal and disembodied soul does not seem to be a part of either of these ancient cultures. This concept of the soul appears to enter Greek thinking through contact with Eastern, shamanistic, and perhaps Thracian cultural sources[145], and it's found explicitly in the Pythagoreans and Orphic cults as well as the writings of Empedocles. Development of this concept culminated in Plato's philosophical treatment of mind and soul, which has been so influential ever since.

Plato's conceptualization of the soul is not a single unified presentation. His descriptions change over time, from one dialogue to another. Generalizing, though, we can say that the soul (as *psyche* is often translated) is more important than the body, is the main entity associated with selfhood, is that aspect of us that must strive for the good and is in closer contact with the Forms, is immaterial, and is immortal. We can also say that the soul has a triune aspect (while still being a single thing) consisting of a lower appetitive and "animal" nature, an emotional ("spirited force") nature, and a higher rational nature (*nous*). The *nous* also plays an important role in Aristotle's treatment of *psyche*. The analytic treatment offered by Aristotle is performed in terms of a matter/form dichotomy similar to his analysis of other phenomena. The form taken on by the matter that allows it to function as a living being is the soul. Soul, in this view, is a property rather than a thing, though the distinction becomes difficult to maintain consistently throughout the discourse. In addition, the rational mind, or *nous*, assumes a different status from all the other functions (which we share with animals) of the soul and has several anomalous characteristics such as separability from the body, imperishability, and connection with divinity.

Notions concerning mind and soul also developed in a variety of cultures far from the ancient eastern Mediterranean. In many primitive cultures, these notions were based less on philosophy and more on myth and ritual. "In regard to death, the rites are all the more complex because there is not only a 'natural phenomenon' (life—or the soul—leaving the body) but also a change in both ontological and social status; the dead person has to undergo certain ordeals that concern his own destiny in the afterlife, but he must also be recognized by the community of the dead and accepted among them."[146]   In India, on the other hand, several sophisticated philosophical traditions grew out of the primary revelation offered by the Vedic scriptures. Within one of these philosophical systems, the *Vaisesika*, soul (or self) and mind are two of "the nine substances which comprise all corporeal and incorporeal things. The existence of soul is inferred from the fact that consciousness cannot be a property of the body, the sense-organs, or the mind. Though the soul is all-pervading, its life of knowing, feeling, and willing resides only where the body is. The plurality of souls is inferred from their difference in status and their variety of conditions. Each soul experiences the consequences of its own deeds..."[147]   Although a descriptive explanation of the soul isn't given in the *Bardo Thodol* of Tibetan Buddhism, instead an elaborate description of the experiences and activities of the soul after death is offered. If the true knowledge of the Void is not grasped at this key time, the soul wanders through a dreamlike state until coming back to the material world through rebirth. This more mystical interpretation is found frequently throughout world cultures. To offer just one more example, the

Persian Sufi Ibn 'Arabi wrote "'I shall return in the end to the beginning, just as in describing a circle the leg of the compass returns to the beginning when it reaches its end. Thus is the end of life bound up with its beginning and its prenatal eternity fuses with the eternity after death. Existence is only transient [...], but there is a lasting enduring vision.'[...] Thus man is dualized by the limits of his consciousness, he consists of an earthly and a transcendent mode of being, and it is the eternal, supralunar aspect of his twofold unity which, according to the religions, survives a man's death; for the most part we remain unconscious or barely conscious of this transcendent aspect, and often it therefore seems nonexistent to us, although the mystics have experienced its reality time and time again and have gained awareness of it as the cause of spirit in themselves."[148]

A number of different conceptualizations of mind and soul have also developed within Christianity. The early Church inherited all the differing views of soul in the Jewish and Greek cultures that formed its developmental matrix. The ideas of Plato were especially influential, as were those of the neo-Platonists such as Plotinus. Other intellectual and spiritual currents that affected the early Church included the Stoics, Epicureans, and Gnostic sects. Within Christianity, the highly original formulations of St. Paul concerning the relationships of body (*soma*), mind (*nous*) and heart (*kardia*), soul (*psyche*), flesh (*sarx*), and spirit (*pneuma*) are both important and also ambiguous. All of these heterogeneous influences lead to a variety of views expressed by the early Fathers of the Church, culminating in a coherent synthesis by St. Augustine. "...the soul is not the whole human, it is his better part. Perhaps, he said, one should rather declare with St. Paul that the soul is the inner human, and the body the outer human [...] the Holy Spirit's gift to humans alone in virtue of their having *rational* souls [...] The rational soul comes to live in a human in a second sense of life; it is not the sense of life conveyed by the identification of soul with the principles of self-movement and alteration [...] in humans alone the ensouled and inspirited 'parts' cohere in one nature. In virtue of its having (or being) a rational soul, God breathes in a spirit which the human can either accept or reject—in accepting, he becomes a new being, and in rejecting he becomes like a beast..."[149] Much later, St. Thomas Aquinas would revisit the issues of mind and soul from the standpoint of Aristotelianism and develop a sophisticated version of how they relate in Christian thought. He starts from Aristotle's metaphysics of matter and form, wherein these must be united in actual objects, like a human body. "According to Aquinas, however, there was one form capable of existing without the body of which it is the form, namely the human soul."[150] Some aspects of the soul, such as the senses, require a body, but the intellect is considered part of the immaterial soul alone. "...the intellect is a power of the soul. It is not identical with the soul; the soul has other powers too, such as the senses and the powers of nutrition [...] Aquinas thinks of the mind as consisting not just of intellect but of intellect plus will [...] he regards thought as an activity which has no bodily organ. Because the activity does not involve the body, he goes on to say that the power, which is the source of the activity, must belong to the soul."[151] Aquinas then works out as coherently as possible how the immaterial and immortal soul can subsist without its body after death.

There seems to be tension between this idea of soul and mind as immaterial and the earlier ideas of St. Paul that implied a greater degree of corporeality in the spiritual substance even while castigating the body/flesh as typically construed. This issue of corporeality appears in a variety of contexts. For example, in the philosophy of Schelling we find a kind of spiritualized materialism that yields ideas of mind and soul requiring some sort of bodily existence, though not the mundane one that we usually enjoy. "Schelling rejected a purely idealistic interpretation of being, with its contempt for physical existence, while at the same time he criticized the materialism of his time for its purely abstract notion of spirit [...] It was Schelling's fundamental belief that there can be no spirit without a body, just as matter cannot exist without an inner life spirit...."[152]   In contrast to these tendencies to attribute some sort of corporeality to our spiritual dimensions, Descartes instead pushes the categorization into material and immaterial aspects to a new extreme. The material body is simply matter in extension behaving according to mechanistic laws, so that the older functions of the soul as vital force and living breath are no longer invoked. Meanwhile, the functions of the rational soul are detached from their older context and redefined as the immaterial mind. Each of these two categories is separately subject to its own reasoning process and epistemological analysis in Descartes's work, but this does not necessarily imply that he believed they were truly separate and separable things. "In terms of the whole person, the mind is intermingled with (*permixtio*) the whole body as its (the person's) own extension. According to the order of reasons as developed through the *Meditations*, clear and distinct knowledge of the essence of mind and the essence of body reveal a *real* distinction between the two, such that they can be conceived as existing independently of each other. But according to the order of essences, clear and distinct knowledge of the whole person reveals that the whole mind and whole body are related as interdependent parts that contribute to a functionally greater whole."[153]

Complementarity and a broader view

We've now examined mind from three points of view, namely the neuroscientific, philosophical/analytic, and philosophical/religious/mystical. These views don't appear coherent with each other. What are we to make of all this? Is the soul (or even the mind) a chimera or sloppy use of language? Or is there more than a single valid way to understand these things and a kind of coherence to be found if we employ complementary perspectives? I believe that humans can in fact be seen as spiritual beings in the light of modern neuro-scientific knowledge. This latter knowledge is acquired under the usual conditions typical of all mundane knowledge: it's based on replicable data that can be verified objectively and interpreted coherently, consistent with any other knowledge we have in the natural sciences and expressible in language that anyone (in principle, at least) can understand. Knowledge of humans as spiritual beings is acquired under totally different (and sometimes mutually exclusive) conditions, which may sometimes involve a component of private first-hand experience and may be incommunicable using normal language and may also depend on a particular cultural context. The validity of the neuroscientific knowledge is virtually undeniable, and I have already offered arguments in favor of the validity of the complementary knowledge concerning the spiritual dimensions of human existence. Here I would like to suggest that both of these kinds of knowledge are needed to formulate a complete conception of mind, and that some of the mysteries we have encountered can be approached productively by this route.

To proceed further down this route, I wish to suggest a few more specific complementary pictures to compare with the mundane view of mind. In this mundane view, only matter is real. Processes that take place are considered to be governed by laws determining how matter behaves and have no further significance. But process can just as well be considered as primary and matter as secondary; there is no compelling metaphysical reason to do otherwise. Processes are what is real in this view and the matter they instantiate into acquires its reality from partaking in the process. There is nothing antiscientific about this assertion, since it's consistent with our present views of both quantum field theory and self-organization (N. B. "consistent with" *not* "entailed by"). If we consider processes as real in this way, then the mind can be associated with these processes of the brain. This view, in which process has an ontological status of its own (we might phrase this by saying that we consider process as object) implies the reality of mind and can also be extended to include the traditional functions of soul (again as process). The complementarity of process as primary and matter as primary involves us in no contradictions unless some process violates laws governing matter. A more difficult question is whether a process can exist independently of any matter instantiating it (we'll discuss this soon). Finally, there is the question of whether we have really gained anything here, or whether we are simply using alternative words to render the same incomprehensible mystery. The two small gains that I see are (1) that we are forced by complementarity to take both mind and brain seriously in an antireductionist schema and (2) that the view of mind as a process and process as a thing might allow us to devise (and even suggest to us) novel relationships between mind and brain that more conventional views would not allow or suggest.

Can a process exist without the matter that the process occurs in? To explore this odd question, let's broaden it further by identifying the process as information, since information specifies a process, a form, a structure, or anything else. Information content is generally embedded in some material manifestation, for example the paint on a canvas specifies the information content of a painting. But, this information can be encoded in a set of numbers in a digital image file. This file still has a material manifestation of some sort, perhaps a set of magnetic domain directions or the voltage values in a set of logic circuits. The information content itself is an abstraction independent of each of these material manifestations in some sense. So, does this information exist independently of these particulars in the material world? This is a metaphysical question, and I would argue that we can answer yes in a complementary viewpoint to that of materialism (where we would need to answer no). If we assign this ontologically existent status to information, then the processes, minds, and souls that correspond to specified sets of information are able to subsist (to use the old-fashioned word) independently of their material instantiations. This strange doctrine contradicts no known objective fact. Indeed, it probably *can't* contradict any such facts and hence might be attacked as unfalsifiable. But this attack can at most disqualify our contention as science, and there's no claim here that it is science (also note the extreme violation of Occam's Razor). The important question is whether the idea of abstract information as an existing nonmaterial thing is somehow useful to deepen our understanding, whether it adds value to our discourse. Before we address that question, note in passing that this idea is rather similar to certain established mystical worldviews. An example is the version of Qabala advanced by many 19th and 20th century mystics, in which the Sephiroth called "Yesod is that subtle basis upon which the physical world is based [...] an omnipresent and all-permeating fluid or medium of extremely subtle matter; substance in a highly tenuous state [...] which is the model upon which the physical world is built."[154]   In my analogical version of this Qabalistic idea, Yesod is where the information resides (although "where" is a misleading word since space and time are part of the physical world and themselves grounded in Yesod).

So how might this idea of information as a nonmaterial existent, which has no apparent scientific content, be valuable? First, let's reemphasize that this ontology is complementary to materialism (which we affirm to be true) and not a replacement for or alternative to materialism. Second, let's reaffirm the validity of the experiential knowledge attested to by many people (saints, mystics, Sufis, Zen practitioners, initiates of the Orphic Mysteries, and so on) that also goes well beyond scientific content and that warrants a spiritual understanding of nature in general and mind in particular. If our goal is to make a coherent worldview that accommodates this knowledge, then the ontology that we're suggesting here has some advantages. One advantage is that the claimed nonmaterial order of reality is not totally disconnected from the material world; the forms, structures, and processes do ultimately instantiate (in part, at least) to the material forms, structures, and processes that we observe, and this imposes some constraints on what might happen even in the nonmaterial realm. A theory of the soul that blatantly contradicts the existence of the brain (and the things we know about it) is untenable. In fact, we might explicitly make contact at this juncture with one of the mainstream schools in philosophy of mind, namely functionalism. The functionalist concept of mind as organizational structure rather than implementation is very much in accord with the ideas presented here, but no longer tied to the need for some kind (any kind) of material implementation. When the functionalist argument arrives at a point where mind as a nonmaterial category is admitted even within an otherwise materialist ontology, it arrives at a dead end. In the view presented here, the emphasis on the one known implementation (the brain) provides us with opportunities to vet our proposals, look for and eliminate contradictions, and invite creative novel interpretations of the mind/brain relationship. Additionally, we have a more effective bridge between the spiritual and mundane modalities of human nature. Extending our previous analogy with contemporary Qabalistic thought, we might say that just as mind is existent in Yesod and existent in the world as instantiated by the brain in matter, this same mind is also related to a higher spiritual Self which is associated in the Qabala with Tiphareth and is the subject of much commentary in Hindu psychology, for example (where it is clearly differentiated from the personal ego).

The simplest interpretation of the foregoing idea would assign a straightforward one-to-one correspondence between any given realized neuronal configuration and some sort of eidolon of this configuration reflected in Yesod (or wherever) as immaterial information. This simple interpretation has some problems, though, and may not be the best way to envision the situation. Our neuronal configurations change with time. Our moods alter, our memories fail, injuries or lesions affect personality, developmental changes make us different people over the years (as do experiences), psychoactive drugs affect our brains and hence our minds, we have religious conversions, we learn new skills, we die, we're born. What is stable and constant in all of this? What is the self? I believe that we need an extended concept of self that includes all of this collectively. A person's entire life, all of their experiences, all of their conscious and unconscious mentality, the entire arc of their material existence from before birth to after death, the memories of their friends and loved ones as well as their own memories, their effects on the world through their interactions in an ever-expanding web; all these things are a person's self. A person's brain is the central locus of all this where the experiences are known and stored and expressed, at a certain moment in time in the manner appropriate to that moment. But events and moments have no meaning in that which is the foundation of the material world as it manifests itself in time (e.g. as previously identified with Yesod). If the mind is existing there, the mind encompasses this extended concept of the person even as the mind also instantiates itself in time by means of the brain. Can the mind then exist without the brain? Certainly in the metaphorical sense I've just described we would need to say yes. A person whose memories, cognition, and sense of self have been corroded by brain lesions is a great tragedy, and we have little reason to believe that their minds are somehow preserved in time as they had once been but located in some immaterial place. But the person is not gone if this person is the sum total of their personhood extended throughout their entire life, and this is the complementary view of person, self, and mind that I am proposing as a serious alternative to a static mundane view. The tragedy, in the example I've used, becomes a part of the person's selfhood and by implication a part of their not-time-bound mind. Although I've developed this idea of an extended self as a web of relations in time and space by starting with the concept of immaterial information, the two concepts are actually independent and can be developed separately; the latter isn't needed to find value in the former. As just one more example, the mundane view of a depressed person is that he or she has some brain deficiency that can be cured with a drug. In the proposed complementary view, the person has an entire life history consisting of experiences and physiological states, with one important experience being an encounter with the drug that alters his or her subsequent experiences, the sum total of which (both before and after) comprise the mind of this person. While this example doesn't explicitly make contact with the sacred, I'd argue that to do so is easier in the complementary description. To make a more explicit connection, recall the often-repeated message from mystics that our true Self exists outside time in Eternity. For this Self, our manifestation in time and the material world must be as I've described it, collective and whole.

Turning to another topic, the reasoning faculty of mind has played a special role in many philosophies, especially in the Western traditions, for millennia. In some cases, the reasoning faculty alone is equated to the mind, and it's also sometimes considered to be a power that transcends bodily functioning. Modern neuroscience offers convincing evidence that this latter contention is not correct, and I believe that a better way to view the mind includes a broader array of attributes, including emotions, perception, memory, will, and self-awareness, as well as ratiocination. Reason is special for many thinkers because we share all these other attributes, except reason, with animals; I don't find this very persuasive, because I believe the differences are a matter of degree and not categorical differences. However, reason still maintains a unique and important role in the scheme of things, because reason allows us to make contact with universals in a demonstrable and comprehensible fashion. This quality is what so impressed Plato, Pythagoras, and Kant about mathematics. That nature behaves in a rational way has convinced many thinkers that nature has a mind or else was shaped by a (divine) mind. That our minds can apprehend this rationality in nature is proof, to some, that we too have a share of the divine. In the mundane view, selection pressures have merely shaped the brain's functioning to include reasoning/language as extra survival tools, mathematics is an invented symbol manipulation game, and the fact that nature exhibits mathematical regularities is due to a combination of cleverness and coincidence. Actually, it's not entirely obvious that extreme forms of these opinions are tenable even in a mundane world. However, the overall attitude is defensible and the fundamental ideas can be considered true, within their horizon, in the complementarity framework we've elaborated. But it's also true, within this complementarity framework, that for a chunk of matter (the brain) to accomplish these tasks is at the very least amazing, is surely highly significant, and is arguably miraculous. Nor is it any less miraculous that we perceive the world, know ourselves, and love. To equate mind with reason is a mistake that was shared by both Cognitive Science and Thomas Aquinas, but an attribute we can share with both machines and angels certainly deserves the central position it occupies in both the mundane and the sacred views of mind.

Lastly, let's briefly look at one more issue, namely the spirituality of corporeal existence. There are a number of religious traditions that regard the body (or some aspect of the body) as inferior to a postulated immaterial mode of being (e.g. the soul is considered to be imprisoned in a tomblike body, trapped in the material world estranged from its true celestial and divine home). In contrast, other traditions assert that matter itself partakes of the divine spirit and that all orders of reality from highest to lowest are indissolubly linked, different aspects of the same divine ground of being. In this latter view, we may coherently refer to the consciousness of matter, though matter is obviously not conscious in the conventional sense of the word. Matter charged with such a spiritual essence may then manifest this quality by means of the forms it acquires, and one of these forms with particular importance here is the human brain. Phrased inelegantly, the brain is a way for spirit to manifest itself in matter and hence know itself explicitly. This theme can be developed in a manner that's independent of any reference to anything immaterial, leaving only matter with existence. But the properties of matter in this context are radically different from the properties of matter in a mundane view. No contradictions need arise, however; these two views are complementary. Whether we have knowledge of the spiritual qualities of matter and see these qualities in the development of mind and brain, or not, depends solely on the conditions under which we acquire our knowledge of matter. We have here an illustration of the ability of the complementarity framework to make distinctions and yield insights: I would argue that everything we know about the brain is more consistent with this idea of spiritualized matter than with the contrasting view of an incorporeal spirit trapped in an evil shell of matter.

To briefly sum up, mind is a part of both mundane and sacred worlds. In a mundane world, mind is a by-product of the activity of the brain. The truth of this statement is virtually undeniable, but the limits to the range of validity of its truth are shown by the philosophical difficulty of specifying just how mind fits into this mundane world. In a sacred world, mind is an aspect of a spiritual Self (or soul) employed to comprehend and experience existence. This statement is equally true within its own range of validity, the limits of which are suggested by the absence of any mention of the brain. The value of the complementarity framework in understanding the mind lies in the explicit connection it makes between brain and soul, and the tools offered for analysis of how they are connected.

An iterative self-reference problem

There is a logical issue lurking in our treatment thus far. The basic problem is this: We are attempting to use the complementarity framework as an analytical tool to better understand the nature of mind. Recall, however, that the development and justification of the complementarity framework was based in large part on the fundamentally necessary presence of a conscious knowing agent. Clearly, this conscious knower can be and must be identified with a mind. Therefore, our analytic tool appears to presuppose the existence of that which it analyzes the existence of.

This problem is reminiscent of Hoffding's point concerning the subject/object relationship, namely that every object is the object of a knowing subject and yet this same subject is also an object of awareness, resulting in an infinite regress. The present version of the problem may be even worse than Hoffding's rendition, since there is no leading term so the regress is infinite in both directions, while at the same time each term of the regress is equivalent to all the others affording no opportunity to approximate an escape. Some relief may be afforded by considering another's mind rather than one's own mind, but the fundamental problem remains. The champion of the mundane view may argue that here lies the fallacy in the entire discourse; let's re-objectify the mind and study it in a purely "scientific" manner, thus getting rid of these paradoxical issues, the appearance of which demonstrate the falsity of this line of investigation from the start. This argument does not go through, however, because the mundane knowledge must still be limited by a selection process that makes it incomplete (even if self-consistent within its own realm). That was the very point that Hoffding was making. We confront an irremediable break in the continuity of our apprehension of Being. The *need* for complementarity is once again reinforced by the iterative self-reference issues that emerge here. And the problem is as acute for the champion of the sacred view as for the mundane.

But what about the implications of this problem for the justification of the complementarity framework itself? A premise can't be used to justify the self-same premise. This fault in the logic used herein must be acknowledged. On the other hand, self-reference problems are hardly unique to the present discourse. The sort of limits we are facing here inevitably turn up in any logical system as it attempts to encompass more and more of what can be imagined until the categories of the system include the system as a category, from the Cretan liar to Godel's theorem. In one sense, our problem here is not quite as severe as in some other cases, because the whole point of complementarity in the present context has been to suggest that understanding all aspects of nature is subject to limits. It is perhaps ironic that our setting of limits to understanding is now subjected itself to the very kind of limits that it posits as inescapable, but this certainly doesn't *weaken* the conclusion that these limits to understanding exist. Indeed, the realization that this problem must be confronted invites us to continue contemplating the issues involved and may lead to yet deeper understanding of mind/brain/soul.

12. A New View of Old Issues

We've now looked in some detail at two issues and the manner in which complementarity may offer new insights into understanding them. In this chapter, a variety of similar kinds of issues are examined, though only quite briefly, in order to get a sense of how complementarity might be more broadly useful. The background material offered is necessarily limited, and our ambition restricted to merely indicating what an analysis would look like and include. A deep and extensive analysis is beyond our scope here. Enough is said, hopefully, to demonstrate the range and power of complementarity as a methodological tool to extend our understanding of important problems.

Natural Theology and Theology of Nature

The term "natural theology" is sometimes used primarily in conjunction with efforts to prove (or at least offer evidence of) the existence of God, but that's not how I will use the term here. I will discuss these kinds of divine existence arguments separately in later sections, and employ "natural theology" in conformance with another common usage of the term that is more broadly conceived. By natural theology, I mean here a kind of discourse "which affirms that at least something of God can be known from the study of nature."[155] In other words, if we presuppose the existence of God based on either faith, gnosis, or some other grounds, we can ask what attributes God might have. Natural theology is an attempt to answer this question by making inferences from our observations of the natural world, to learn something about the creator by looking at the creation. There is a strong and well-known tradition of this type of activity, especially in the early history of the rise of science in Europe and in much of the theological literature of the same time period.

Theology of nature is a somewhat different kind of activity. A theology of nature "starts from a religious tradition based on religious experience and historical revelation. But it holds that some traditional doctrines need to be reformulated in the light of current science."[156] In this case, assertions about both God and nature are not limited to results from the scientific study of nature, but both kinds of assertions are also informed by such scientific study. Quite a bit of the contemporary writing on issues of science and religion, especially by Christian writers, falls into this category. This body of work ranges from the poetic and metaphorical to the philosophically rigorous, and includes examples of scientific visions inspiring novel theological perspectives, scientific facts constraining the use of outmoded theological doctrines, and theological interpretations of natural activities in terms of ultimate meaning and purpose.

Both of these activities can be considered within the complementarity framework, but the application of complementarity is rather different in each case. In the case of natural theology, the starting point becomes a scientific view of nature that we informally think of as a set of facts concerning nature. These facts are no different from a mundane or from a sacred perspective. If we draw some further implications from these facts (concerning God, for example), then such implications are either true or not true depending on a set of presuppositions that we hold having nothing to do with the observation of nature. To illustrate this point with a specific example, recall

Kepler's discoveries about the mathematical structure of the planetary orbits. Kepler was both a devout Lutheran and also deeply imbued in the Pythagorean tradition of mystical and divinely ordained numerical relationships as the foundation of the world. His discoveries were for him (and for anyone who joins him in this) a wonderful new insight into the mysterious workings of God in the creation and sustaining of the universe. The same mathematical relationships have an entirely different meaning for a committed atheist, representing nothing more than the success of human ingenuity in describing compactly the information content of the empirical measurements of planetary positions. These two assertions sound contradictory, but is either Kepler or the atheist actually wrong? They are both employing the same set of well-established facts about nature, and neither has violated any logical rules. I would therefore argue that these two assertions (and the world-views they exemplify) are complementary and can both be true. The non-overlapping conditions under which the knowledge is acquired are the very presuppositions mentioned previously.

This argument goes through in a similar way for any comparable example of natural theology, but there are still several unresolved issues to consider. I have argued that both assertions can be true, but this argument doesn't address the question of whether both assertions must be true. What are the warrants for the truth claims of any such interpretation, and how can they be evaluated? What are the limits to what one can assert given a particular set of facts about nature? From the perspective of science, what is the added value or the danger of engaging in natural theology? From the perspective of theology, a much narrower vision emerges from any kind of natural theology approach than from the wider purview of revelation, scripture, etc. On the other hand, the constraints imposed by natural theology on our conceptions of divinity also result in more consistency among different groups of people with different religious traditions. All of these issues would need to be analyzed more carefully in a thorough consideration of natural theology, but the fundamental insight offered by complementarity is a useful starting point.

Much of this discussion applies equally well to the theology of nature. The major difference is that because the inherently religious presuppositions are more narrowly focused, both the advantages (a more ecumenical view) and the disadvantages (a more conceptually impoverished view) from the theological perspective become more muted. The role of complementarity also becomes subtly altered, because in this case the complementarity is between entirely different visions of nature (with potentially radical levels of meaning included) going well beyond the facts themselves, even if such facts are respected. The fundamental similarity is in the conditions of acquiring knowledge of nature lying in a presupposed knowledge of divine attributes that don't depend on natural knowledge. But because the presuppositions are more extensive and the complementarities extend so far outside the mere interpretation of natural contingent facts themselves, the evaluation of truth claims in a theology of nature becomes even more problematic. Radically differing theologies (think of Christianity and Hinduism, for example) might both be made consistent with the same set of facts about nature and still make extremely different claims about some truly fundamental points. In what sense are such sharply differing claims complementary? This kind of important question needs to be taken seriously if complementarity is to be genuinely valuable as a tool in the consideration of theologies of nature.

Synchronicity

Synchronicity is a term coined by C. G. Jung to describe situations in which some event occurring in the objective external world has a correlation with another psychological event that occurs in the internal world of a person. The particular nature of the correlation is that the external event is meaningfully related to the internal event. There is no causal connection between these two events, so they are related only by the meaningfulness they have for the person who experiences them. Let's illustrate this idea with a fictional example: Suppose you are in the midst of some personal crisis that can only be resolved by an act of courage on your part, but you have not come to realize that this is needed to solve your problem. As you are walking along ruminating on this unresolved issue in your personal life, you cross paths with a lion that has escaped from the zoo. Lions, you realize in a flash of insight, are an archetypal symbol of courage, and at the same time you realize what is needed to resolve your problem. The appearance of the lion at that point in your life, when it had this important meaning for you, is an example of synchronicity. In general, "two essential features of every synchronistic experience emerge: First, an objective event or series of outer events meaningfully relates to a subjective psychological state (dream, fantasy, or feeling) [...] The second essential feature is the lack of causal connection between the outer event and the subjective inner state."[157]   These sorts of situations apparently happened to Jung's therapeutic analysis clients so frequently that he coined the term synchronicity to describe them and incorporated this concept into his overall theory of depth psychology as an important component, indicative of a more general pattern of acausal connections in the world. Jungian analysts and their clients, as well as many other people, continue to experience striking instances of synchronicity in their lives.

Critics of this concept dismiss the idea that anything interesting or special is going on here, and claim that these events are obviously merely coincidences. You and the lion both had to be somewhere, and you both happened to be in the same place at the same time. The lion crossed many people's paths. The fact that you happened to derive some meaning from your encounter is irrelevant to how we should believe the world is ordered. Indeed, an acausal relationship between two events near each other in time is more-or-less the definition of a coincidence. Is a synchronicity really something beyond a simple coincidence that should have a different status in our understanding of the world, or is it just another word for the same thing?

The answer to this question turns entirely on the significance that we attach to meaning, because meaningfulness is the defining characteristic of a synchronicity. If all meaning is drained from the world, as it certainly must be within an entirely mundane view, then the meaning you attach to some event is merely an epiphenomenon signifying nothing. Coincidence is then all that remains. Conditions under which knowledge is acquired of whether the event in question is synchronistic are clearly defined in this case and virtually guarantee that the answer is no. But are there an alternative set of conditions under which synchronicity can also be considered a valid concept within some complementary standpoint defined by such conditions? We come here once again to the crucial role played by the knowing subject. At one pole, the subject stands detached (as much as possible) from the rest of the world and objectifies it into the mundane view. At the opposite pole, the subject is totally immersed in relationships with the world (again as much as possible) and these relationships define both the subject and the world. Standing at this pole, the world is unavoidably imbued with meaning defined by these relationships. The mutually exclusive nature of detachment and relationship are pretty clear, and so these two poles epitomize very generally the knowledge acquisition conditions we've been discussing, from which more particular conditions in specific cases can be identified. Our previous epistemological arguments have sought to show that *both* of these poles are necessary to attain an exhaustive understanding of a situation, that either one alone is incomplete. From this proposition, we conclude that meaning must have an important significance, since without it we are missing one of the perspectives demanded by complementarity for an exhaustive understanding, and hence that synchronicity is a valid category. The conclusion of this line of argumentation is that while some event may indeed be a perfectly good example of mundane coincidence, this does not stop the same event from being an equally good example of synchronicity.

While the view that denies synchronicity is clearly identified with what I generally term mundane, it's not as clear that the view that validates synchronicity is the same as what I generally term sacred. Both cases (synchronicity and the sacred) demand the genuine significance of meaning, but they are not necessarily logically identical. On the other hand, it is the case that Jung and his school regard synchronicity as a part of the process of achieving psychological integration and wholeness (individuation), and that they quite generally regard this process as intimately related to spiritual development. Jung and his school also regard an occurrence of synchronicity as an encounter of the personal ego with something that transcends it, and although they try to treat this "something" scientifically they unavoidably tend to describe this transcendent "self" using language that is similar to a religious language. For example, Jung speaks of "a new approach that opens the door to a direct, spontaneous religious encounter with nature, by which Jung meant a direct, spontaneous encounter with the living symbols of the unconscious psyche"[158] For these reasons, I believe that only an exceedingly narrow gulf separates the complementary view in which synchronicity is found from a sacred apprehension of the world.

Design Arguments

Proofs of the existence of God have a long history and a variety of forms, but they never turn out to be truly compelling in any logical sense. The reasons for this state of affairs have been thoroughly explored by many philosophers and theologians, but these discussions aren't entirely relevant to the concerns I'm about to take up. Although some design arguments have in the past been presented as putative proofs of God's existence, the modern versions are more modestly offered as "best explanations" rather than genuine "proofs" of divine origin or intervention in the universe. Best explanations of what? That's the interesting question.

We will briefly examine two very different cases. First, we will look at the so-called "intelligent design" movement that argues in favor of the following premise: biological systems, especially at the cellular level, have a kind of irreducible complexity that is unexplainable based on purely naturalistic grounds. This concept, often abbreviated as ID, garnered a lot of notoriety in the popular media in 2005 due to court cases and to essays by high Church officials. The second case we will consider is often referred to as the "fine tuning" of the physical constants in nature. The claim here is that only a very improbable set of values for these fundamental constants could have allowed life to exist in the universe, and that this set of values is just what actually occurs. The argument is that this fact offers evidence of design in how the universe is constructed.

It's difficult to discuss ID in a dispassionate manner because of all the ideological baggage it carries. There is reason to believe that some proponents of ID are using it as a tool to promote a broader social, cultural, and political agenda (an agenda that I, frankly, disagree with). I will ignore this completely for the purposes of the present discussion and treat the arguments at face value, as subtle philosophical points. The arguments consist of two related contentions.

One contention is that there is no justification on *a priori* grounds for the exclusion of "nonnaturalistic theories"[159] from science. The second contention is that empirical observations of the world in conjunction with mathematical probability arguments and related failures of naturalistic explanations for these observations lead to the conclusion that a nonnaturalistic theory, which they term design, is the best explanation for the origins of life and many of the particular features of living organisms. Put more bluntly, Darwin's mechanism of evolution by natural selection doesn't explain the data, so an intelligent designer must be invoked to do the job. The first contention is interesting from a philosophy of science viewpoint. Although it is neither irrational nor indefensible, I ultimately find it unpersuasive because it contradicts a presently existing cultural understanding of science that is virtually universal. In any event, the first contention becomes moot if the second contention is wrong, and the mainstream scientific community has presented a very strong critique indicating that the irreducible complexity argument does not hold. But if ID fails as science, does it hold any interest for us outside of science? Before addressing this question, let's look at fine tuning.

By fundamental constants of nature, we mean numerical values for quantities such as the speed of light, the charge of the electron, Newton's gravitational constant, Planck's constant, and so on. Since these constants determine important aspects of the physical world like the strength of gravity and the relationship of mass to energy, their values actually govern, in some sense, what kind of physical reality exists. A different set of values for these constants would result in a radically different kind of universe. We have experimentally measured the values of these constants to astonishing degrees of precision. But why do they have the particular values that they have? No one knows. Perhaps some day we will have enough scientific knowledge to understand why the fundamental constants have these values, but for now they remain contingent facts of nature having no explanation. A few decades ago, it was noted that many of these constants need to be close to their actual values in order to have a universe with properties compatible with the presence of life. For example, if the cosmological constant were not extremely close to its actual value, the universe would be either very short-lived or else composed of very thinly spread hydrogen atoms. Neither of these scenarios could support the existence of life. Another example concerns the values of the constants that govern the strength of the electromagnetic and strong nuclear forces. If either of these constants were slightly different, then the nucleosynthesis processes in stars would produce very little carbon or oxygen. Many more such examples can be cited, and the bottom line is that the entire collection of fundamental constants seems to have values that are fine tuned so as to result in a universe that can indeed sustain life. The statement that this must be so is sometimes called the Anthropic Principle. But what is the real significance of all this?

There is a school of thought that regards these remarkable facts as evidence for a design argument. The argument is that no result as *a priori* improbable as this could be a mere accident or coincidence, "that God designed our universe with the ultimate emergence of self-reflective consciousness as its intention."[160] Even if this is not proof of God's existence, the argument goes, at least it is evidence; invoking an intentional act of God is the best explanation for these facts. Critics, not surprisingly, disagree. They point out that this contention has no explanatory power at all, and that what we need is a more comprehensive physical theory in which the values of the constants are an understandable and calculable result. Also, there is a class of cosmological theories (inflationary models) that postulate a vast number of non-interacting universes, so in this kind of model we merely happen to inhabit the one that's inhabitable, which is no mystery. These rival schools continue to argue with each other, but it seems to me that both sides are not thinking about the issues properly.

Let's start in the mundane world. Any sort of inference that a design partisan might make from the kind of observations we've been discussing is clearly meaningless within a mundane apprehension of nature, since the material content of the world and its regulated behavior is all that this view encompasses. Now, an interesting point here is that the marshaling of these kinds of observations is intended by the design partisans just precisely for the purpose of attacking the mundane view, to try to prove (or at least indicate) that this view is untenable. But this attempt is almost surely doomed to failure, because the person living in a mundane world has ruled out (not as a matter of logic but by virtue of horizon) the very foundations of the inference. Only an absolutely irrefutable demonstration, a truly compelling argument that tolerates no rebuttal, could possibly succeed in convincing a person living in a mundane world of the factuality of divine design, and if it did then this person's world would be shattered. But this irrefutable and compelling argument, of course, can't be made, leaving the denizens of the mundane world safe within their presuppositions and leaving the design partisans puzzled about why their mounting piles of evidence fail to convince.

Now let's move into a complementary apprehension of the world as sacred. In this case, we need no sophisticated scientific evidence for the presence of divinity in the created world. Breathing is enough, even for those who don't know the role of oxygen in metabolism. In breathing the air, we *know* the miraculous aspect of the presence of air for us to breath, of our own presence being here to breath it. If you don't know this, then no amount of rational argumentation will impart the knowledge; if you do know it, then no rational argumentation is needed. In this world, observations indicating design have an acknowledged home. Is the air here so I can live by breathing it? Of course it is. The foregoing statement doesn't contradict our scientific understanding of the evolutionary development of ecosystems, because the question isn't a question about the historical mechanisms leading to the situation. The question is an existential question about my situation in the here-and-now, and the answer is perfectly valid. The exquisitely complex biological structures of flagellae and eyes, the astonishingly well-wrought values of the fundamental constants, and the air I'm breathing are all the earmarks of a transcendent presence operating in the world. You can call this presence a designer if you wish, although to me the word "designer" conjures up anthropomorphic images of God as a tinkering engineer, and I find such images singularly unattractive. One might even contend, within the context of a sacred worldview, that to employ these things in some sort of evidentiary way, as part of an existence argument, is at best tasteless and at worst verges on idolatry.

The most interesting aspect of this analysis concerns the logical status of these various observations (e.g. "irreducible complexity" and "fine tuning"). All of this knowledge was generated within a mundane worldview. On the one hand, this makes even clearer why the design argument *qua* argument is ineffective, because it attempts to employ inherently mundane knowledge to erode the mundane worldview. On the other hand, my previous comment that this knowledge represents "the earmarks of a transcendent presence" within the sacred worldview must mean that some mundane knowledge has seeped into our sacred world. But this doesn't cause any problems if we are cautious, in fact something like this must always occur. The oil for the sacred lamps must have been extracted by some fairly mundane process. The difference here is the highly abstract level of the mundanely generated knowledge being incorporated into the complementary sacred view grounded in a much more direct contact with nature. The situation is actually quite interesting because it illustrates nicely the following important point: the object of knowledge in our complementary views is always the same entity apprehended in different ways.

The Challenge from Traditional Sciences

The cultural authority of modern science was challenged during the twentieth century by a loosely associated group of thinkers who might be labeled the "traditionalist school." They rejected not only science (as understood today) but essentially all of modernism, which they would replace with a world-view more consistent with that of traditional premodern cultures. These traditionalists should be distinguished from fundamentalists, who they consider to be a symptom of modernity no less than the scientists. Instead, they consider the heart of the wisdom contained in these premodern cultures to lie in their esoteric traditions. These esoteric traditions themselves are embedded within the context of the traditional religions (Sufism as a part of Islam, for example), and they are the carriers of the great spiritual, metaphysical, and cosmological truths known to mankind for millennia. Modernism, in their view, is causing the death of these eternal truths by cutting us off from their sources, and it must therefore be resisted. Among the major writers associated with this school are Guenon, Schuon, Burckhardt, and Nasr. Influences of this school are also found in the work of scholars such as Coomaraswamy and Huston Smith.

Perhaps the major central contention of the traditionalists is that there are metaphysical truths that are undeniable, that predate mankind, and that trump any mere empirical knowledge we might obtain. "These principles, by reason of their very universality, are essentially inherent in human intelligence at its most profound [...] with the aid of supernatural elements that an authentic and complete spiritual tradition alone can supply."[161]   Although the major world religions appear on the surface to be quite different in their beliefs and orthodoxies, according to the traditional school this superficial appearance is misleading due to the exoteric nature of these beliefs.   The esoteric truths underlying all of these differing religious beliefs reveal a fundamental unity that they refer to by various names such as *sophia perennis*. If modern science suggests contradictory conclusions such as a heliocentric planetary system or the evolutionary development of species, these conclusions can be discounted as mere appearance. The traditionalists offer an array of sophisticated philosophical arguments in favor of their contentions, many of them familiar to us by now. For example, what warrant do the claims of modern science have if the human reason that generates these claims has itself no warrant in a world-view of scientific materialism? We cannot recount and critically examine all of these arguments here. Anyone who actually understands the genuine basis upon which the modern scientific world-view is erected, however, must surely be left with nagging doubts about the universality of the traditionalist truth claims for a view that denies so many things we have good reason to believe. On the other hand, many of the points made by the traditional school are compelling and important, even if they seem alien in the modern world.

I am suggesting that complementarity offers a useful tool with which to approach the traditionalist critique of modernism and try to salvage a world-view in which the sacred knowledge they are preserving can be understood without doing irremediable damage to the mundane knowledge that we have worked hard to obtain. The program is obvious enough: identify aspects of the traditionalist and materialist world-views that constitute common topics of discourse; analyze the differing claims in terms of the different conditions under which knowledge of the topic is acquired, and use this analysis to determine which claims are complementary and which are contradictory; in the latter case, allegiance must be given to one claim or the other, based on whatever criteria are considered decisive; in the cases where the claims are complementary, we may believe both without any logical problems. We can't carry out this ambitious program for the entire wide-ranging critique of the traditionalists in conjunction with the enormous corpus of work that makes up the modern scientific picture, but my intuition is that almost all of the truly crucial differences will turn out to be complementary.

In lieu of a thorough analysis of all the issues, let's take a brief look at one important point of controversy to demonstrate how such a program might work. Is a human being fundamentally an animal species that has resulted from natural evolutionary processes, or a microcosm of the divine order in a sacred cosmos, an *imago dei*? These propositions, the former a claim of scientific materialism and the latter from the traditionalist school, certainly appear to be antithetical. But let's consider the conditions under which we can know either of these things. The first proposition is embedded in a large and complex ontology that excludes *a priori* the sources of knowledge that might lead to the second proposition. We first restrict our attention to empirical observations from the fossil record, from genetic studies of existing species, and so on. We then coherently integrate all of this empirical knowledge into a common context, find that this context is reasonably complete and self-consistent, and formulate a narrative structure around it. From all this we conclude the truth of the first proposition, and if we regard this truth as exhaustive then we deny the truth of the second proposition as contradicting the first. Now, the second proposition is also embedded in a large and complex ontology that is completely different in its presuppositions. Species in this case are material manifestations of preexisting Forms (similar to those of Plato) and all of material existence is connected to unseen orders of reality that are apprehended directly through an experiential gnosis that doesn't depend on the senses. The mechanism by which these species become materially manifest in time is virtually irrelevant, and sensory evidence is not regulative to the most important aspects of reality. The experiential gnosis is developed within and reinforced by the esoteric practice and wisdom of nearly every culture on earth, contact with which constitutes another condition for knowing. As we look more closely, it becomes clear that the conditions under which knowledge of what it means to be human are almost entirely different in the two cases. They are both self-consistent and coherent within their own horizons, and the limited areas of overlap need not result in any contradictions. While a more comprehensive analysis might be nice, I believe we already have done enough to show that these two propositions are indeed complementary. We may believe both of these ideas concerning humanity, one mundane and the other sacred; to do so is neither incoherent nor problematic. The only restriction would be on holding both ideas, along with everything each represents, fully and simultaneously.

Environmental Ethics

The issues in this section are rather different from most of the work we are doing. Generally, the central task has been to define more precisely the conditions under which knowledge is acquired and use this epistemological analysis to delimit the region of validity for some particular worldview. In the present case, we will consider the question of a sacramental aspect to nature as a kind of background to the main issue at stake here. This main issue is an issue of ethics and values, not of knowledge claims *per se*, but the knowledge claims are an integral part of the process by which we think about the ethical concerns under consideration. Actually, various kinds of ethical

concerns have already been inherently linked to the epistemological conclusions we've drawn, but we have not examined
them in detail. Here, we will explicitly but briefly look at this
one specific example to illustrate how the ethical implications
of our work might be developed. The example we will consider
is environmental degradation, more particularly the kind of
extreme global environmental degradation that the power
of modern technology has made possible and that presently
threatens the global ecosystem with such a major collapse that
human civilization, perhaps even human survival itself, is at risk.
I will not bother documenting the factual grounds for making
the foregoing dramatic statement; appropriate documentation
is widely and easily available, and the interested reader should
consult it.

The question I will take up here is not whether this
environmental degradation is occurring, because it obviously
is. The relevant question here is why this manifestly suicidal set
of actions continues to occur. The irrationality of destroying
your own means of sustenance might alone seem enough to
stop it, but as Cardinal Newman asserted well over a century
ago, reason is an extremely weak bulwark against the greed and
passions of the human race. The good Cardinal would employ
religion rather than reason to combat human evil, but the role
of religion and spirituality in promoting a wise and healthy
environmental ethic has been ambiguous and problematic. In
the last few decades, a large literature addressing this topic has
developed. For example, one widely influential thesis holds that
the biblical passage bestowing dominance to humans over all the
rest of nature is an important source of the present ecological
crisis. "Christianity [...] not only established a dualism of
man and nature but also insisted that it is God's will that man
exploit nature for his proper ends [...] Christianity made it
possible to exploit nature in a mood of indifference..."[162]
This thesis has been contested, but even many of those who

believe it point to an alternative vision and tradition within Christianity that is exemplified by the nature mystics and saints like St. Francis and Hildegard. The emphasis here is on the goodness of creation, the sacramental aspect of nature implied by incarnational thinking, and the role of humans as the good stewards of creation (the keepers of the Garden) rather than its exploiters. "The world is understood as organic to God, not as a mere product of his will. This means also that the world cannot be conceived in narrowly anthropocentric terms, as if it were provided solely for man's exploitation."[163] There is little evidence, however, that this sort of thinking has taken hold in contemporary culture, either because the other paradigm is too strongly rooted or because a secular capitalist ideology of exploitation is really in control. In any event, other spiritual traditions also carry ecological overtones. Buddhist ideas of compassion for all beings and mindfulness, for example, can be relevant to environmental problems. "...the environmental crisis for humans is at heart a spiritual crisis. Great strength of spirit is needed to respond to the overwhelming dimensions of environmental destruction. All religious traditions have the capacity to awaken and inspire the human spirit. I believe that three principles of Buddhist everyday practice and philosophy can make a major contribution to this awakening."[164] In China, the Buddhist and Taoist traditions have sometimes maintained a sense of relationship to nature and also balance with nature. "China is still an inherently sacred landscape [...] In particular the scores of sacred mountains across China offer islands of environmental protection which have almost all vanished elsewhere [...] there are over a hundred such sacred mountains, often only known to be sacred in their immediate environment but protected from exploitation by the devotion of local families and villages..."[165] The possible examples here are too numerous to discuss in any detail, but the direct contact with nature found in shamanic and Native American traditions, the special role of nature in Japanese Shinto beliefs, and the role of humans as God's viceregent on earth found in Islam are all worth mentioning. Of course, humans have always altered their environmental surroundings in the course of making a living, and not always for the good. For example, a great deal of deforestation has been documented even in the history of China, despite Taoist writers extolling the virtues of being in balance with nature. "....the enlightened memorials to the emperor on the need for the conservation of resources are in themselves clear evidence of the follies that have already been committed."[166] But we presently don't have even an ideological counterweight to slow unbridled exploitation and environmental degradation, and we presently have an unprecedented ability to engage in it. Religious and spiritual traditions appear to vary in their attitudes toward and ability to curtail these assaults on nature. Is there anything within a mundane view of nature to limit our greed and exercise of power?

There is nothing in the mundane view that confers an intrinsic worth on nature, but even within the mundane view there are limits to our actions grounded in self-interest. By self-interest, I am including both personal egotistic self-interest but also including a broader conception of self-interest that includes all of humanity. In this latter sense, the "self" means "we ourselves," the human race, as opposed to other species of animals or plants, other members of the global environment. For example, even if we don't care whether many trees die for *their* sakes, we should care because they supply the oxygen we need to breath. In addition, acquisition of knowledge within the mundane view frequently informs us about environmental threats that we otherwise would not be aware of. Field biologists tell us when a species is declining by measuring it. Global warming is a subtle effect; to discover it required precise measurements of the global temperatures and atmospheric carbon dioxide levels, along with very sophisticated mathematical modeling of the global climate. The role of the sciences in environmental degradation has been complex, both positive and negative, but it's no accident that many scientists have been in the forefront of the fight against the assault on the global ecosystem. The role of spiritual traditions has likewise been complex, both positive and negative. A sacred view of nature might foster a relationship in which nature is cherished, but there have also been world-denying tendencies in some religions that leave them free to abuse nature without any sense of wrongdoing. So, there is a complicated interplay between the ways in which to apprehend reality and the ethical stance we adopt in regard to nature.

There is also another problem. We face yet an additional logical complication in considering the relationship of humans to the rest of nature and what this relationship might imply. Humans are a part of nature. This seems a reasonable proposition in either a mundane or a sacred approach. But, this fact implies that the artifacts produced by humans are also a part of nature. There is nothing inherently unnatural about a substance (some plastic, for example) synthesized by the humans who are themselves natural. If many such substances poison the environment, so be it; this too is a natural process. After all, when microorganisms first produced oxygen eons ago, the oxygen was a deadly poison to existing life forms, and so they became extinct. There have been many mass extinction events in the long history of life on earth, and if we are precipitating one of them now then we are simply another part of the natural process. Why should we care? The earth may seem like a desolate wasteland to us (if we survive at all) in a hundred years, but our present ecosystem is pretty nasty-looking to organisms killed by oxygen. Things change in nature, and we have no reason to privilege the present ecosystem over whatever it evolves into subject to our actions. Arguments of this sort are made both by some greedy industrialists and developers justifying their rape of the earth and by some deep ecologists who claim to eschew anthropocentrism. Obviously (to me, anyway), this argument does not pass the common-sense test. But where is the flaw in the argument? The initial premise seems to be correct; we don't want to separate humans from the rest of nature in either the mundane view (where there are no possible grounds for doing so) or in the sacred view (where we might have grounds for it, but would pay the price of severing the relationship that we are trying to maintain between humanity and nature). The flaw in the argument now becomes apparent, because it is this very relationship, within the context of a sacred worldview, that does privilege the present state of earth's ecosystem.

Both the privileged status of nature as it presently exists and the relationship we enter into with it are not matters for logical deduction. These things are direct intuitive apprehensions of nature. A beautiful creek with clear clean water running through a verdant wood is better than a flow of toxic sludge running through the stumps of dead and decaying trees. That is not a logical deduction; that is just something I *know*. And my knowledge of this trumps any line of reasoning to the contrary. Yes, it's true that change is inevitable in nature and that the presence of humans will inevitably bring about some of this change. These truisms, however, do not imply that massively destructive change caused by humans is ok. The difference between acceptable change and destruction is a matter of balance, and the person with a sacred relationship to the rest of nature has the ability to make good judgments about whether balance is being maintained. Another dimension of the sacred view of nature is that humans play a special role in creation. Humans are a part of nature, but not *just* a part of nature, comparable to worms or grasses. By virtue of their self-conscious awareness and their rationality, the very things that give them so much power over the rest of nature, humans bear a unique obligation and responsibility to the rest of nature (an idea found in many traditional religions). This is the ethical duty toward the environment that follows from living in a sacred world. From this special status of humans in the world as sacred, the special (and hence privileged) status of the global ecosystem in its present state follows by implication, but the implication here is completely mutual since the sacredness of the world flows from transcendent sources and confers upon humans their uniqueness and their obligations.

Interestingly, these conclusions based on a sacred view are identical to the conclusions that follow from a mundane view. The processes by which we arrive at the conclusions are complementary, but in this case the conclusions themselves turn out to be the same. While it's true that nature has no intrinsic worth itself within a mundane view, even the instrumental concept of nature as useful to humans leads us to conclude that destroying our global ecosystem is not really a good idea. The "reasoning" that misses this point may appear cogent to a certain school of economists, but a genuine rational analysis reveals that the conceptualization of self-interest employed there is pathologically narrow. While people might convince themselves that destroying one part of the environment or another is a satisfactory price to pay in order to gain some sort of personal reward, even this kind of self-justification fails when the cumulative effect is leading toward global collapse. And even if we don't have any ethical obligations towards nature in a mundane world, we still have ethical obligations toward other humans, including those who will live a few generations from now. Global environmental destruction is just as wrong in a mundane world as it is in a sacred world.

It seems to me a remarkably powerful result that starting from either of two complementary ontologies we arrive at the same ethical conclusions regarding the treatment of the environment. Perhaps some day this confluence of wisdom and utilitarianism will be enough to outweigh the greed and stupidity that govern our treatment of the environment at the present time.

Alchemy

The topic of alchemy is complex and multifaceted, approachable through the lens of history, psychology, comparative religion, chemistry, literature, or occultism. Alchemical traditions have existed in China, India, Egypt, the Islamic world, medieval and renaissance Europe, and into modern times; these traditions have all borrowed from each other as well as evolving independently. Although some people still think of alchemy as pseudoscience, quackery, and charlatanism (and some alchemists have perhaps fit this caricature), no serious scholar thinks this way about the topic. Scholars do, however, have controversies and disputes about whether alchemy is primarily concerned with chemical reactions, psychological states, cosmological paradigms, or spiritual growth. I will sidestep these historical and ideological issues here, accepting the presence of all these factors in alchemy and not worrying about the relative importance of each one. I will take for my paradigmatic case the alchemy of renaissance Europe that was related to many broader currents of hermetic thought, and this sort of generic "hermetic alchemy" will serve as a model (a kind of ideal type) upon which to base my comments here.

The alchemical world view is distinctively non-modern, and by implication not a mundane view. This is not to say, however, that the alchemist doesn't share some important aspects of the modern approach to nature. In particular, both alchemy and modern materialism share a similar idea of matter as behaving in lawful and regular ways, and each one constitutes a search for these regularities in order to understand matter more fully. Moreover, both approaches have a strong empirical component. The empiricism of the modern approach is self-evident, but the relationship of alchemy to metal smelting and working, to dye and perfume making, to glass making and blowing, and to other practical craft lore surely indicates that empirical data have always been important in alchemical work. The important discoveries of alchemists (gunpowder being perhaps the most spectacular) interpretable in terms of modern chemistry are well known, and these also demonstrate a strong empirical component in their work. But the crucial difference between modern empirical approaches to nature and the traditional alchemical approach turns on the role of the practitioner who makes the empirical observation. The alchemist is not a detached observer but rather a participant in the processes taking place. For example, Fludd writes "that the man which is partaker of this divine Agent, and can firmly unite it unto his own spirit, may do wonders."[167]   One of the standard formulations of the so-called spiritual alchemy holds that the purification of the metals in the alchemical process is and must be accompanied by the purification of the practitioner's soul. The efficacy of the operations on matter then depend importantly on the internal state of the operator.

The alchemists also had a fundamentally different understanding of the nature of matter than the modern understanding. For the alchemists, matter is a living substance (hylozoism) that grows and transforms. "Wherefore there is no corporeall thing which hath not a spirit lying hid in it, as also a life [...] For not only that hath life which moves and stirres [...] but also all corporeall and substantiall things."[168]   Such hylozoism is not unique to alchemy. Similar ideas are found in many cultures and probably entered alchemical thought naturally through those cultures, but in alchemy this conception of matter as being alive and spiritually charged became part of the theoretical structure for the study of matter through operative work. The joining of two substances to create something new was envisioned in sexual terms, for example, and the transmutation of metals was thought to be a hastening of the natural gestation and growth process occurring deep within the earth.

Finally, the idea of nature held by the alchemist is grounded in a broader conceptualization of reality that transcends nature itself. As the Emerald Tablet, one of the primary documents of the alchemical tradition, states: "That which is above is like to that which is below, and that which is below is like to that which is above." The laws and regularities governing the behavior of matter are a reflection of spiritual laws and regularities that might be understood better by examining matter, but the reverse was at least equally true. Only by achieving a more transcendent understanding of broader aspects of reality could the alchemist come to understand the processes going on in the alembic. As a relatively recent commentator has phrased it, "alchemical cosmology is essentially a doctrine of being, an ontology. The metallurgical symbol is not merely a makeshift, an approximate description of inward processes; like every true symbol, it is a kind of revelation."[169]

Alchemy presents us with a particularly interesting case, because the alchemists were keen observers of the behavior of matter in their workshops and they made significant contributions to our knowledge, yet they clearly held a sacred view of nature rather than the mundane view that dominates modernism. Observing the same chemical reactions, they beheld a different world. How can we evaluate these differences? Within the framework of complementarity, our analysis centers on the conditions under which knowledge is acquired, i.e. the conditions under which the observations are made. The first obvious difference between a mundane view and the alchemical approach is the kinds of questions being asked. In some cases the questions are primarily empirical, and in these cases the alchemical and mundane worlds virtually coincide. But a question that presupposes a teleological intentionality or something akin to sexual desire in the reacting materials is inconsistent with the mundane view and yet still consistent with the observations. Likewise for a question that presupposes atomic states with quantitatively measurable electronegativities, in this case consistent with a mundane view. The two questions are both compatible with the observations, but they are logically incompatible with each other. We can legitimately ask either question, but not both simultaneously, which satisfies the criterion for complementarity. Another condition under which knowledge is acquired that must be specified here is the state of consciousness of the practitioner. We can't know this with any certainty for the alchemists, but judging from the alchemical imagery that we have available it's reasonable to infer that many of these alchemists were in a different mental state than a contemporary individual performing the same chemical reaction. The literature associated with the so-called "spiritual alchemy" also strongly suggests that these workers were in some sort of altered state of consciousness related to mystical union with the divine, rather than the sober and rational state that the average present-day chemist would be in while working. Once again the interpretation of the observed reaction would be much different in the two cases, the two interpretations being complementary and both correct within their appropriate contexts. Yet another condition under which knowledge is acquired is the cosmological presuppositions held by the practitioner, presuppositions that must be greatly influenced by cultural factors as well as by individual circumstances. Matter that reflects some aspect of a divine attribute must be seen differently from matter that constitutes the sum total of existence in itself, even though that matter is "doing the same thing." The complementarity here obviously arises from incompatible presuppositions, as illustrated in the example. All of these conditions (questions asked, state of consciousness, cosmological presuppositions) are interlinked with each other to create two different worlds, one sacred (held by our hypothetical alchemist) and the other mundane (which we more typically inhabit).

We must be careful in our analysis, however, because not all statements are necessarily complementary. Statements that make knowledge claims within the horizon of the mundane view might well be contradictory to other statements within the same horizon. For example, the idea that metals gestate within the earth and that iron will eventually evolve into gold through such a process directly contradicts the more stable binding energy of iron and the origin of heavy elements in supernova explosions. I consider this mundane knowledge to be reliable enough to rule out the competing alchemical "model" and can't consider the two ideas to be complementary statements that are both true. This is not a failure of complementarity. I believe, instead, that examples like this demonstrate the power and usefulness of complementarity as a methodology that can make distinctions based on epistemological analyses. Meanwhile, the sacredness of gold as a solar symbol could easily be recovered without the accretion of a scientifically incorrect model of its origins. In chapter 10, alternatively, I suggested a possible kind of sacred symbolic interpretation of the formation of heavy elements that is consistent with scientific facts and would be complementary with the mundane view. As far as alchemy is concerned, it would require a great deal of work (involving history, philosophy, and mystical thought) to do a careful and thorough analysis and devise a coherent sacred view of nature that is based on alchemical premises and consistent with contemporary knowledge. Such a project may or may not be possible, and it's probably not worth the effort it would require, but the possibility is intriguing. A more delimited analysis of particular texts and ideas, performed along these lines, is more practical and might be worthwhile as an example of how a sacred world-view can partially be constructed.

Romantic Science

The somewhat unsatisfactory title of this section refers to a collection of alternative approaches to the study of nature that are not all identical to each other but all share the trait of rejecting the purely mechanistic approach that characterized Enlightenment science. Thinkers in this tradition span a range of interests in philosophy and science, including Schelling, Humboldt, Goethe, Oersted, and Faraday. The *naturphilosophie* movement is the most well-defined expression of these ideas, but the general approach both predated and outlasted *naturphilosophie*. Some of the characteristics of this approach are an emphasis on holism, a sense of nature as organic, and a greater valuation of direct experience of nature over abstraction from nature. A number of important discoveries in the sciences, such as the field concept, appear to have been influenced by this approach. Many of the ideas and results that were proposed by its practitioners, however, are conventionally dismissed as being wrong or unproductive. While some of these ideas and results may genuinely be wrong in the sense of contradicting something we know to be true on legitimate grounds, I believe that many of them may instead be complementary and hence true within their proper horizon. Carefully examining even a fraction of the large corpus of such work would be a huge task, so instead I will merely take a single example to analyze by way of illustrating my point.

The example I will use concerns the methodology of Goethe. Unlike the standard methodology of modern science, which makes abstractions of the properties of things and then discovers relationships between the abstractions, Goethe made the actual observations of the things the foundation of his method. "Goethe held that the investigator should begin not with the pole of speculative thinking or first principles, but with observation, and should take care not to replace observations with abstractions."[170]   His method is what we would now call phenomenological, i.e. he observes ever more deeply and penetrates to a level of understanding that's not apparent on the surface. From his observations he develops a kind of archetypal comprehension of the phenomenon that's based not on abstraction but on intuitive insight. "One begins with ordinary 'empirical phenomena,' the simple observations any attentive observer might make. From these one can rise to a higher awareness by varying the conditions under which the phenomenon appears. By doing so, the essential preconditions for its appearance become apparent. Such instances he termed 'scientific phenomena.' [...] but he sought a higher level of recognition—what he called 'pure phenomena' or, later, 'archetypal phenomena' (*Urphanomen*)."[171] The goal here is a kind of deep internal insight into nature, an experiential knowing that cannot be easily communicated to another person. In a way, the method is also the result, and Goethe's writings are meant to show the way for another person to employ the method and achieve their own insight. This again is very unlike our conventional methods today, which are meant to be unambiguously communicable in words and equations. The result can be a kind of knowledge available in Goethe's romantic science that can't be found from doing conventional science. "As thinking comes alive in nature, and nature comes alive in the activity of thinking, knowledge of the world and knowledge of the self unite at a higher level..."[172]. At the same time, the kinds of things we learn in conventional science can't be discovered using Goethe's method either.

These points can be illustrated by briefly comparing Goethe's theory of colors with Newton's theory. Goethe's theory of colors is often said to be wrong, and his attempt to supplant Newton's ideas about color with a rival paradigm is dismissed as a misguided dead end in the history of ideas. That view is reasonable within the context of modern science and its goals, but as we have seen Goethe's goals are quite different. To dismiss a rival of the Enlightenment program simply because it's not the Enlightenment program begs the question. I would contend that the theory of colors developed by Goethe is complementary to the conventional ideas because the conditions under which Goethe wishes us to acquire our knowledge about colors is entirely different from those of Newton and his successors. "Goethe looked first at the colors that are formed when the prism is used in light in the natural environment, instead of the restricted and artificial environment that he felt Newton had selected as the experimental basis of his approach. By doing this, Goethe recognized that the phenomenon of prismatic colors depended on a boundary between light and dark regions. Far from the colors somehow being already *contained* in light, for Goethe, they *came into being* out of a relationship between light and darkness [...] the prism was a complicating factor, and so to understand the arising of colors, he looked for the more simple cases, which meant [...] only light and darkness [...] Thus, it was in the natural environment that Goethe first recognized the primal phenomenon of color [...] he was in a position to see how the colors change from one to another as conditions change [...] these shifts were at the root of more complex phenomena such as the prismatic colors. One result is that a dynamic wholeness is perceived in the prismatic colors—a wholeness totally lacking in Newton's account."[173] We see here that even the actual experimental conditions need to be different, but more profoundly the mental state of the practitioner is totally different. The analytical and quantity-minded mentality of Newton is not compatible with the intuitively observant mentality of Goethe. The same person might affect both of these internal mental conditions, but not at the same time. In consequence, the kind of knowledge we discover is radically different in the two cases. Newton discovers relationships between the pieces of the light (the colors), while Goethe discovers the wholeness of the light that discloses itself to us in varying ways (the colors). These are not contradictory conclusions in the sense that one must be wrong because the other is right. They can each be right under the appropriate conditions and we have found that these conditions do not mutually overlap. So, they are both right in the sense of complementarity.

How does this relate to our more general discussion of the complementarity between the sacred and mundane? Although Goethe's method is empirical and scientific, his version of romantic science is not consistent with a mundane world, either ontologically or methodologically. His method requires us to enter into a relationship, as it were, with the phenomenon. Nature must have a quality of organic wholeness that's antithetical to materialism in order for Goethe's science to make sense. If we look carefully at the conditions required for Goethe's way of doing science, we see that many of these conditions are quite similar to those required for the sacred apprehension of nature. "Goethe claims that intuitive perception of a whole is a valid scientific method. Through perception of archetypal images (*Urbilder*), we participate in the creative process within nature's innermost spiritual core [...] The parallel Goethe draws between contemplative observation of nature and the raising of oneself through religion into a higher region shows just how highly Goethe valued his method [...] Here, natural cognition—the perception of nature's archetypal images—is addressed directly as divine revelation."[174] That such a sacred apprehension of nature is here presented within the context of a systematic methodology for the study of nature makes this an extremely interesting case. A more extensive examination of Goethe's phenomenological way of doing science (and of romantic science in general), employing the framework of complementarity as an analytical tool, could be exceptionally valuable.

Epilogue

So, we have arrived at the end of our intellectual journey. What have we accomplished after this strenuous effort?

At a minimum, I believe that we have at least managed to clarify some of the relevant issues involved. My hope, however, is that two further things have been accomplished: I hope that rabid enemies of either of the two worldviews I have been defending (the mundane and the sacred) will perhaps manage to open their minds to possibilities that have eluded them heretofore. And I hope that some who have tried to embrace both of these worldviews, but felt perplexed by the apparent contradiction they presented, will find something of value in the analysis I present here. I know that I found value in it.

And among experts in the field of science/religion studies, I have a further hope: My hope here is that I have contributed a worthwhile methodological tool to the discipline. There are a host of knotty epistemological problems left to address, and many details left to deal with in the applications of these ideas to particular problems, but despite all this undone work I think that a fruitful start has been made here. My final hope is that someone will eventually travel further down this path.

Before ending, let us summarize the central argument one last time. We start with the premise that there is an inherent inseparability between the knowing subject and the known world in our apprehension of nature. This is so because we only know the world through our experience of it. (These points are in analogy to Bohr's conclusions regarding the situation in quantum physics, necessitated in that case by the non-continuous dynamical changes at the microscopic level.) This experience is conditioned (in some sense at least) by our minds (broadly considered, including language, culture, neurophysiology, and so on), limiting our available concepts. Due to these limitations, our understanding of nature is only meaningful when the concepts that we use and the conditions under which we use them are carefully examined. (Once again, an insight inspired by Bohr's work, in which case the "conditions" are purely experimental because the phenomena are restricted to physics and the empirical sciences.) The results of such a careful examination yield clusters of concepts (which we can conveniently label "nature as sacred" and "nature as mundane") that appear to be contrary descriptions but that in fact are indicative of differing and mutually non-overlapping conditions of knowing; both descriptions are necessary in order to have an exhaustive description of nature itself. This final conclusion is what we have termed complementarity. The sacred apprehension of nature is complementary to the mundane apprehension. And yet once again, the terminology and the argument itself were inspired by Bohr, who concluded that such a viewpoint was necessary in quantum physics and that it explained the otherwise mysterious results found there. The novel contribution in the present work has been to expand the "conditions of knowing" beyond the restriction to empirical phenomena in order to deal effectively with explicitly spiritual matters. The conclusion I have tried to defend based on this reasoning is that nature is *both* sacred *and* mundane, with no necessary contradiction between these two contentions if the analysis is performed carefully.

Because I believe that ideas of this sort are only valuable in practice rather than as abstractions, I have engaged in the use of complementarity (as developed herein) to explore a number of longstanding problems in the science/religion arena: the problem of cosmology and origins; the mind/brain problem; the theology of nature; the design argument; and a number of others. While I obviously haven't solved any of these intractable issues, I do think that complementarity has offered some fresh insights into the issues; I invite the reader to make their own judgments whether this is the case.

I have an acquaintance who is a Native American healer/ shaman and is also deeply involved in societal affairs; he once made the statement that "I live in two worlds." I believe he meant that literally, and in the sense that I have described here. To a limited extent, I know what he means, as I at least dimly apprehend both the sacred and mundane aspects of the world. But I also have a passion for coherence in my view of reality, and this motivated the project that evolved into the present book. As I said above, my hope is that others who share my predilections will find some value in it.

Endnotes

1. E. MacKinnon, 1993
2. K. H. Reich, 2002
3. D. M. MacKay, 1957
4. MacKay, 1974, p. 226
5. H. J. Folse, 1985
6. J. Faye, 1991
7. K. J. Sharpe, 1991
8. P. P. Duce, 1996
9. I. G. Barbour, 1990
10. P. Alexander, 1956
11. H. A. Bedau, 1974
12. F. Watts, 1998
13. H. Walach and N. von Stillfried, 2011
14. J. E. Loder and W. J. Neidhardt, 1996
15. J. Honner, 1985
16. A. Weber, 1896, p. 23
17. R. G. Collingwood, 1945, p. 36
18. G. de Santillana, 1961, p. 121
19. G. Lloyd and N. Sivin, 2002, p. 214
20. ibid., p. 221-222
21. J. Needham, 1981, p. 14
22. Tu Wei-Ming, in L. S. Rouner, 1984, p. 115
23. Lloyd and Sivin, p. 198
24. Needham, p.15
25. M. Iqbal, 2002, p. 3
26. M. G. S. Hodgson, 1974, p. 431
27. H. R. Turner, 1997, p. 17 & 31

28. Iqbal, p. 71-72
29. ibid, p. 92-93
30. Turner, p. 47
31. ibid., p. 191
32. Hodgson, p. 200
33. Iqbal, p. 36
34. R. Vitzthum, 1995, p. 229-231
35. S. Radhakrishnan and C. A. Moore, 1957, p. 227
36. J. K. Feibleman, 1970, p. 40
37. M. Artigas, 2000, p. 303
38. Vitzthum, p. 104
39. J. K. Feibleman, p. 151
40. TwoBears, in J. E. Carroll et al., 1997, p. 159-160
41. Collingwood, p. 3
42. F. Bauerschmidt, 2003, p. 106-107
43. McDaniel, in Carroll et al., p. 109
44. G. W. Russell, 1918, p. 170-171
45. Zajonc, in D. Seamon and A. Zajonc, 1998, p. 27
46. M. Eliade, 1959, p. 3-4
47. S. H. Nasr, 1996, p. 15
48. Bauerschmidt, p. 109-110
49. R. Sheldrake, 1991, p. 221
50. Eliade, p. 20
51. Nasr, p. 60-63
52. Harris, in G. F. McLean, 1978, p. 31-37
53. Nasr, p. 58
54. Meier, in J. Campbell, 1960, p. 167-168
55. Westfall, in J. Torrance, 1992, p. 78
56. Toulmin, in Rouner, p. 29
57. E. J. Dijksterhuis, 1950, p. 361-362
58. R. Popkin, 2003, p. 123
59. Dijksterhuis, p. 310
60. Dijksterhuis, p. 323
61. Westfall, in D. C. Lindberg and R. L. Numbers, 1986, p. 221-222
62. Dijksterhuis, p. 425
63. ibid., p. 414-415
64. Ashworth, in Lindberg and Numbers, p. 139
65. Dijksterhuis, p. 442
66. Deason, in Lindberg and Numbers, p. 185-186

67. Dijksterhuis, p. 441
68. Popkin, p. 73
69. Westfall, in Torrance, p. 81-82
70. Jacob, in Lindberg and Numbers, p. 249
71. Roger, in ibid., p. 287
72. Hahn, in ibid., p. 270
73. Popkin, p. 219
74. Moore, in Lindberg and Numbers, p. 332
75. Rudwick, in ibid., p. 306
76. ibid., p. 310-311 (emphasis in original)
77. ibid., p. 314
78. Dupree, in Lindberg and Numbers, p. 365
79. N. Bohr, 1963, p. 54
80. Lamprecht, in Y. H. Krikorian, 1944, p. 18
81. Nasr, p. 64
82. Bohr, p. 91
83. ibid., p. 6
84. quoted in R. D. Morrison, 1994, p. 73
85. W. James, 1907, p. 71 (emphasis in original)
86. R. D. Romanyshyn, 1989, p. 115
87. D. T. Suzuki, 1959, p. 331-363
88. ibid., p. 16
89. M. Merleau-Ponty, 1969, p. 27
90. J. D. McCurdy, 1978, p. 70
91. James, p. 133
92. H. Smith, 2001, p. 38
93. W. M. Shea, 1984, p. 25
94. B. Lonergan, 1967, p. 213-214
95. H. Corbin, 1969, p. 181
96. Mahadevan, in McLean, p. 184-186
97. A. Huxley, 1945, p. ix
98. H. Hoffding, 1906, p. 75 (emphasis in original)
99. ibid., p. 79-80
100. ibid., p. 84-85 (emphasis in original)
101. ibid., p. 107-111 (emphasis in original)
102. ibid., p. 113
103. Merleau-Ponty, 1968, p. 137
104. Morrison, p. 28, p. 66, and p. 283
105. Hoffding, p. 82

106. ibid., p. 112-115
107. for a discussion, see Reich, 2002, ch. 5
108. E. Nagel and J. R. Newman, 1958, p. 6
109. Abe, in McLean, p. 168-169
110. Hoffding, p.120
111. ibid., p. 142-144
112. A. Liddle, 1999, p. 44
113. R. A. Alpher and R. Herman, 2001, p. 18
114. N. Austin, 1990. p. 5
115. D. Maclagan, 1977, p. 9
116. L. Stookey, 2004, p. 213
117. ibid., p. 43
118. P. Freund, 1975, p. 36
119. Peters, in R. J. Russell et al., 1988, p. 276
120. Barbour, p. 130
121. McMullin, in Russell et al., p. 56
122. Peters, op. cit., p. 288-291
123. Stoeger, in Russell et al., p. 231
124. Barbour, p. 147
125. U. Goodenough, 1998
126. B. Swimme and T. Berry, 1994
127. A. R. Globus, 1988
128. M. Gonzalez-Wippler, 1974, p. 73
129. Nathan, in G. Underwood, 2001, p. 53
130. J. LeDoux, 2002. p. 58
131. R. Joseph, 1993, p. 345
132. Posner and Rothbart, in C. Koch and J. L. Davis, 1994, p. 197-198
133. Haugeland, in R. Cummins and D. D. Cummins, 2000, p. 48
134. J. Heil, 1998, p. 4-5
135. D. J. Chalmers, 1996, p. 123
136. J. Kim, 1996, p. 211-212
137. adapted from Kim, p. 222
138. Place, in Cummins and Cummins, p. 366
139. Heil, p. 123
140. S. Aurobindo, 2001, p. 332
141. Collingwood, p. 3-4
142. Wei-Ming, in Rouner, p. 114-124
143. Artigas, p. 320

144. P. S. MacDonald, 2003, p. 16
145. ibid.
146. Eliade, p. 185
147. Radhakrishnan and Moore, p. 386
148. Meier, in Campbell, p. 158-159
149. MacDonald, p. 155-156
150. A. Kenny, 1993, p. 25
151. ibid., p. 42
152. Benz, in Campbell, p. 233
153. MacDonald, p. 290-291
154. I. Regardie, 1932, p. 61
155. A. McGrath, 2002, p. 128
156. Barbour, p. 26
157. V. Mansfield, 1995, p. 23-24
158. R. Aziz, 1990, p. 10
159. M. J. Behe et al., 2000, p. 153
160. Gingerich, in P. Kurtz, 2003, p. 56-57
161. T. Burckhardt, 1967, p. 17
162. White, in D. Spring and E. Spring, 1974, p. 24-25
163. Macquarrie, in ibid., p. 45-46
164. Kaza, in J. E. Carroll et al., p. 155
165. Palmer, in D. E. Cooper and J. A. Palmer, p. 27
166. Tuan, in Spring and Spring, p. 105
167. Fludd, in S. J. Linden, 2003, p. 191
168. Paracelsus, in Linden, p. 155
169. Burckhardt, p. 27
170. Zajonc, in D. Seamon and A. Zajonc, 1998, p. 24
171. ibid., p. 25
172. Cottrell, in Seamon and Zajonc, p. 259
173. Bortoft, in ibid., p. 290
174. Heitler, in ibid., p. 65

Bibliography

Alexander, Peter, Complementary Descriptions, Mind **65**, 145, 1956.

Alpher, Ralph A. and Herman, Robert, *Genesis of the Big Bang*, Oxford University Press, Oxford, 2001.

Artigas, Mariano, *The Mind of the Universe*, Templeton Foundation Press, Radnor PA, 2000.

Aurobindo, Sri, *A Greater Psychology* (A. S. Dalal, editor), Tarcher/Putnam, New York, NY, 2001.

Austin, Norman, *Meaning and Being in Myth*, University of Pennsylvania Press, University Park, 1990.

Aziz, Robert, *C. G. Jung's Psychology of Religion and Synchronicity*, State University of New York Press, 1990.

Barbour, Ian G., *Myths, Models, and Paradigms*, Harper & Row Publishers, New York, 1974.

Barbour, Ian G., *Religion in an Age of Science*, HarperCollins Publishers, New York, 1990.

Barnes, M. (editor), *An Ecology of the Spirit*, University Press of America, Inc., Lanham, MD, 1994.

Barrow, J. D., *The Constants of Nature*, Pantheon Books, NY, 2002.

Bauerschmidt, Frederick, *Why the Mystics Matter Now*, Sorin Books, Notre Dame, IN, 2003.

Bedau, Hugo Adam, Complementarity and the Relation Between Science and Religion, Zygon **9**, 202, 1974.

Behe, M. J., Dembski, W. A., and Meyer, S. C., *Science and Evidence for Design in the Universe*, Ignatius Press, San Francisco, 2000.

Bodde, Derk, *Chinese Thought, Society, and Science*, University of Hawaii Press, Honolulu, 1991.

Bohm, David, *Wholeness and the Implicate Order*, Routledge, London, 1980 (reprinted 1988).

Bohr, Niels, *Atomic Theory and the Description of Nature*, The MacMillan Co., New York, NY, 1934.

Bohr, Niels, *Atomic Physics and Human Knowledge*, John Wiley & Sons, Inc., New York, NY, 1958.

Bohr, Niels, *Essays 1958-1962 on Atomic Physics and Human Knowledge*, Wiley, New York, 1963 (reprinted Ox Bow Press, Woodbridge, CN, 1987).

Bohr, Niels, *Niels Bohr Collected Works Vol. 6*, ed. by Jorgen Kalckar, North-Holland Physics Publishing (Elsevier Science Publishers), Amsterdam, Netherlands, 1985.

Burckhardt, Titus, *Alchemy*, Penguin Books Inc., Baltimore, MD, 1971 (reprint; first published in German 1960, translated in 1967 by W. Stoddart).

Burckhardt, Titus (translated and edited by W. Stoddart), *Mirror of the Intellect*, State University of New York Press, Albany, 1987.

Campbell, Joseph (editor), *Spiritual Disciplines*, Bollingen Foundation and Pantheon Books, New York, NY, 1960.

Campbell, Joseph (editor), *Man and Transformation*, Bollingen Foundation and Pantheon Books, New York, NY, 1964.

Campbell, Joseph (editor), *The Mysteries*, Princeton University Press, Princeton, NJ, 1978 (reprinted from 1955 Bollingen Foundation edition).

Capra, Fritjof, *The Tao of Physics*, Shambhala Publications, Inc., Boston, MA, 1975 (reprinted 2000).

Carroll, John E., Brockelman, Paul, and Westfall, Mary (editors), *The Greening of Faith*, University Press of New England, Hanover, NH, 1997.

Chalmers, David J., *The Conscious Mind*, Oxford University Press, Oxford, 1996.

Collingwood, R. G., *The Idea of Nature*, Oxford University Press, Oxford, 1960, reprinted 1967 (originally published by Clarendon Press, 1945).

Cooper, D. E. and Palmer, J. A. (editors), *Spirit of the Environment*, Routledge, NY, 1998.

Corbin, Henry, *Alone with the Alone*, Princeton University Press, Princeton, NJ, 1998 (first published by Flammarion in French, 1958; first English edition 1969).

Cummins, R. and Cummins, D. D. (editors), *Minds, Brains, and Computers:*

the Foundations of Cognitive Science, Blackwell Publishers Ltd., Oxford, UK, 2000.

Dijksterhuis, E. J., *The Mechanization of the World Picture*, translated by C. Dikshoorn, Princeton University Press, Princeton, NJ, 1950 (translated 1961, reprinted 1986).

Duce, Philip P., Complementarity in Perspective, Science & Christian Belief **8**, 145, 1996.

Eliade, Mircea, *Cosmos and History*, Harper & Row, NY, 1959 (originally published in French in 1949, translated and published by Pantheon Books as *The Myth of the Eternal Return* in 1954).

Eliade, Mircea, *The Sacred and the Profane*, Harcourt, Brace, & World, Inc., New York, NY, 1959.

Faye, Jan, *Niels Bohr: his Heritage and Legacy*, Kluwer Academic Publishers, Dordrecht, Netherlands, 1991.

Feibleman, James K., *The New Materialism*, Martinus Nijhoff, The Hague, Netherlands, 1970.

Folse, Henry J., *The Philosophy of Niels Bohr*, North-Holland Physics Publishing (Elsevier Science Publishers), Amsterdam, Netherlands, 1985.

Freund, Philip, *Myths of Creation*, Transatlantic Arts, Inc., Levittown, NY, 1975.

Globus, Alfred R., *Veritism*, The Foundation for Science and Theology, Inc., 1988.

Gonzalez-Wippler, Migene, *A Kabbalah for the Modern World*, Julian Press, Inc., NY, 1974.

Goodenough, Ursula, *The Sacred Depths of Nature*, Oxford University Press, Oxford, 1998.

Granit, Ragnar, *The Purposive Brain*, MIT Press, Cambridge, MA, 1977.

Griffin, David Ray, *Reenchantment Without Supernaturalism*, Cornell University Press, Ithaca, NY, 2001.

Grossenbacher, P. G., *Finding Consciousness in the Brain*, John Benjamins Publishing Co., Philadelphia, PA, 2001.

Harrison, Edward, *Masks of the Universe*, Cambridge University Press, Cambridge, 2003 (first edition published 1985).

Hayek, Friedrich A., *The Sensory Order*, University of Chicago Press, Chicago, 1976 (reprint, originally published 1952).

Heil, J., *Philosophy of Mind*, Routledge, New York, NY, 1998.

Hodgson, Marshall G. S., *The Venture of Islam*, University of Chicago Press, Chicago, 1974.

Hoffding, Harald, *A History of Modern Philosophy*, translated by B. E. Meyer, Danish edition 1895, English translation Macmillan and Co. reprinted Dover Publications, Inc., NY, 1955.

Hoffding, Harald, *The Problems of Philosophy*, translated by Galen M. Fisher with a Preface by William James, The MacMillan Co., New York, NY, 1906.

Honner, John, S.J., Unity-in-Difference: Karl Rahner and Niels Bohr, Theological Studies **46**, 480, 1985.

Huxley, Aldous, *The Perennial Philosophy*, Harper and Brothers Publishers, NY, 1945.

Iqbal, Muzaffar, *Islam and Science*, Ashgate Publishing Limited, Hampshire, England, 2002.

James, William, *Pragmatism*, Longmans, Green and Co., Inc., 1907 (World Publishing Company, 1955; New American Library, Inc., New York, NY, 1974).

Joseph, R., *The Naked Neuron*, Plenum Press, New York, NY, 1993.

Kenny, Anthony, *Aquinas on Mind*, Routledge, New York, NY, 1993.

Kim, Jaegwon, *Philosophy of Mind*, Westview Press, Boulder, CO, 1996.

Krikorian, Yervant H. (editor), *Naturalism and the Human Spirit*, Columbia University Press, New York, NY, 1944.

Koch, C. and Davis, J. L., *Large-Scale Neuronal Theories of the Brain*, MIT Press, Cambridge, MA, 1994.

Kurtz, Paul (editor), *Science and Religion*, Prometheus Books, Amherst, NY, 2003.

LeDoux, Joseph, *Synaptic Self*, Viking Penguin, New York, NY, 2002.

Liddle, Andrew, *An Introduction to Modern Cosmology*, John Wiley & Sons, NY, 1999.

Lindberg, David C., and Numbers, Ronald L. (editors), *God and Nature*, University of California Press, Berkeley, CA, 1986.

Linden, Stanton J. (editor), The *Alchemy Reader*, Cambridge University Press, 2003.

Lloyd, Geoffrey, *Early Greek Science*, W. W. Norton & Co., New York, NY, 1970.

Lloyd, Geoffrey, *Greek Science After Aristotle*, W. W. Norton & Co., New York, NY, 1973.

Lloyd, Geoffrey and Sivin, Nathan, *The Way and the Word*, Yale University Press, New Haven, 2002.

Loder, James E., and Neidhardt, W. Jim, Barth, Bohr, and Dialectic, in *Religion and Science; History, Method, Dialogue*, Routledge, NY, 1996.

Lonergan, B., *Collection: Papers by Bernard Lonergan, S.J.* (ed. by F. E. Crowe), Herder and Herder, NY, 1967.

MacDonald, Paul S., *History of the Concept of Mind*, Ashgate Publishing Company, Burlington, VT, 2003.

MacKay, D. M., Complementary Descriptions, Mind **66**, 390, 1957.

MacKay, D. M., Complementarity in Scientific and Theological Thinking, Zygon **9**, 225, 1974.

MacKinnon, Edward, Complementarity, CTNS Bulletin **13.1**, 12, 1993.

Maclagan, David, *Creation Myths*, Thames and Hudson, London, 1977.

Malin, Shimon, *Nature Loves to Hide*, Oxford University Press, Inc., Oxford, 2001.

Mansfield, Victor, *Synchronicity, Science, and Soul-Making*, Open Court Publishing Company, Peru, IL, 1995.

McCurdy, John Derrickson, *Visionary Appropriation*, Philosophical Library, Inc., New York, NY, 1978.

McGrath, Alister, *The Reenchantment of Nature*, Doubleday, New York, NY, 2002.

McLean, George F. (editor), *Man and Nature*, Oxford University Press, 1978.

Melhado, E. M. and Frangsmyr, T. (editors), *Enlightenment Science in the Romantic Era*, Cambridge University Press, Cambridge, 1992.

Merleau-Ponty, Maurice, *The Visible and the Invisible*, translated by Alphonso Lingis, Northwestern University Press, Evanston, 1968 (original in French, Gallimard, 1964).

Merleau-Ponty, Maurice, *The Essential Writings of Merleau-Ponty*, ed. by Alden L. Fisher, Harcourt, Brace, & World, Inc., New York, USA, 1969.

Milner, A. D. and Rugg, M. D., *The Neuropsychology of Consciousness*, Academic Press, San Diego, CA, 1992.

Morrison, Roy D. II, *Science, Theology, and the Transcendental Horizon*, Scholars Press, Atlanta, 1994.

Nagel, Ernest, and Newman, James R., *Godel's Proof*, New York University Press, 1958.

Nasr, Seyyed Hossein, *Religion and the Order of Nature*, Oxford University Press, Inc., New York, NY, 1996.

Needham, Joseph, *Science in Traditional China*, Harvard University Press, Cambridge, MA, and Chinese University Press, Hong Kong, 1981.

Newman, W. R. and Grafton, A. (editors), *Secrets of Nature*, MIT Press,

Cambridge, MA, 2001.

Omnes, Roland, *The Interpretation of Quantum Mechanics*, Princeton University Press, Princeton, NJ, 1994.

Peters, Ted (editor), *Science and Theology*, Westview Press, Boulder, CO, 1998.

Polkinghorne, John, *One World*, Princeton University Press, Princeton, NJ, 1986.

Popkin, Richard, *The History of Scepticism from Savonarola to Bayle*, Oxford University Press, Oxford, 2003.

Quine, W. V. O., *Word and Object*, Technology Press of MIT and John Wiley & Sons, NY, 1960.

Radhakrishnan, S. and Moore, C. A. (editors), *Sourcebook in Indian Philosophy*, Princeton University Press, Princeton, NJ, 1957.

Raine, D. J. and Thomas, E. G., *An Introduction to the Science of Cosmology*, Institute of Physics Publishing, Bristol, UK, 2001.

Redhead, Michael, *Incompleteness, Nonlocality, and Realism*, Oxford University Press (Clarendon), Oxford, 1987.

Regardie, Israel, *A Garden of Pomegranates*, Llewellyn Publications, Saint Paul, MN, 1970 (first edition 1932, fourth printing 1978).

Reich, K. Helmut, The Relation Between Science and Theology: the Case For Complementarity Revisited, Zygon **25**, 369, 1990.

Reich, K. Helmut, *Developing the Horizons of the Mind*, Cambridge University Press, Cambridge, 2002.

Romanyshyn, Robert D., *Technology as Symptom & Dream*, Routledge, London and New York, 1989.

Rosenzweig, M. R., Leiman, A. L., and Breedlove, S. M., *Biological Psychology*, Sinauer Associates, Inc., 1999.

Rouner, Leroy S. (editor), *On Nature*, University of Notre Dame Press, Notre Dame, 1984.

Rugg, M. D. and Coles, M. G. H., *Electrophysiology of Mind*, Oxford University Press, Oxford, 1995.

Russell, George William (aka AE), *The Candle of Vision*, University Books, New Hyde Park, NY, 1965 (originally published 1918).

Russell, Robert J., Stoeger, William R., S.J., and Coyne, George V., S.J. (editors), *Physics, Philosophy, and Theology*, Vatican Observatory, Vatican City Sate, 1988.

Santillana, Giorgio de, *The Origins of Scientific Thought*, The New American Library, New York, NY, 1961 (originally published by the University of Chicago Press).

Scheibel, A. B. and Wechsler, A. F., *Neurobiology of Higher Cognitive Function*, Guilford Press, New York, NY, 1990.

Schelling, F., *Ideas for a Philosophy of Nature*, translated by E. E. Harris and P. Heath, Cambridge University Press, Cambridge, 1988 (original edition 1797).

Seamon, David and Zajonc, Arthur, *Goethe's Way of Science*, State University of New York Press, 1998.

Sharpe, Kevin J., Relating Science and Theology With Complementarity: a Caution, Zygon **26**, 309, 1991.

Shea, William M., *The Naturalists and the Supernatural*, Mercer University Press, 1984.

Sheldrake, Rupert, *The Rebirth of Nature*, Bantam Books, NY, 1991.

Smart, Ninian, *Dimensions of the Sacred*, University of California Press, Berkeley and Los Angeles, CA, 1996.

Smith, Huston, *Why Religion Matters*, HarperCollins Publishers, NY, 2001.

Spring, David and Spring, Eileen, *Ecology and Religion in History*, Harper & Row, NY, 1974.

Stapp, Henry P., *Mind, Matter, and Quantum Mechanics*, Springer-Verlag, Berlin, 1993.

Stookey, Lorena, *Thematic Guide to World Mythology*, Greenwood Press, Westport, CT, 2004.

Suzuki, Daisetz T., *Zen and Japanese Culture*, Bollingen Foundation, Inc., New York, NY, 1959 (Princeton University Press, Princeton, NJ, 1970; eleventh printing, 1993).

Swimme, B. and Berry, T., *The Universe Story*, HarperSanFrancisco, San Francisco, CA, 1994.

Toolan, David, *At Home in the Cosmos*, Orbis Books, Maryknoll, NY, 2001.

Torrance, John (editor), *The Concept of Nature*, Oxford University Press (Clarendon), Oxford, 1992.

Turner, Howard R., *Science in Medieval Islam*, University of Texas Press, Austin, TX, 1997.

Underwood, G. (editor), *Oxford Guide to the Mind*, Oxford University Press, Oxford, 2001.

Vitzthum, Richard, *Materialism*, Prometheus Books, Amherst, NY, 1995.

Wagner, Steven J. and Warner, Richard (editors), *Naturalism*, University of Notre Dame Press, Notre Dame, 1993.

Walach, Harald and von Stillfried, Nikolaus, Generalized Quantum

Theory—Basic Idea and General Intuition: a Background Story and Overview, Axiomathes **21**, 185 (2011).

Wallace, B. Alan, *The Taboo of Subjectivity*, Oxford University Press, Oxford, 2000.

Watts, Fraser, Science and Theology as Complementary Perspectives, in *Rethinking Theology and Science*, ed. by Gregersen, N. H., and van Huyssteen, J. W., Wm. B. Eerdmans Publishing Co., Grand Rapids, MI, 1998.

Weber, Alfred, *History of Philosophy*, translated by Frank Thilly, Charles Scribner's Sons, New York, 1896.

Whitaker, Andrew, *Einstein, Bohr, and the Quantum Dilemma*, Cambridge University Press, Cambridge, 1996.

www.ingramcontent.com/pod-product-compliance
Lightning Source LLC
Chambersburg PA
CBHW062048270326
41931CB00013B/2984